THIS BOOK BELONGS TO

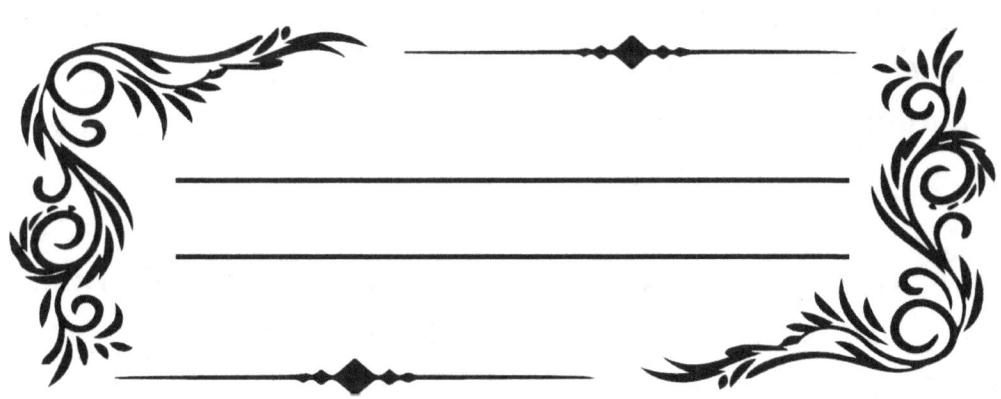

Created & designed by
2 school teachers.

"**Exciting World of ABCs and 123s: A Fun Handwriting and Counting Adventure** is an engaging and educational book designed to introduce preschoolers to the world of handwriting, letter tracing, number recognition, and counting. Through a series of fun and interactive activities, young learners will develop essential skills while exploring the joy of writing and counting in a playful and encouraging environment.

Key Features:

1. **Learn the ABCs**: Through engaging illustrations, children will be introduced to the 26 letters of the alphabet, each accompanied by adorable images that correspond to the letter's sound.
2. **Trace and Practice**: The book features interactive tracing exercises for each letter, providing ample practice for children to perfect their handwriting. Large, clear letter guides ensure easy tracing for little fingers, making learning an enjoyable and rewarding experience.
3. **Counting Fun**: Alongside the alphabet journey, this book will introduce children to counting. Children will explore numbers 1 to 20 engagingly.
4. **Interactive Activities**: Throughout the book, interactive activities like "Guess the Letter" and "Count the Objects" will keep young learners excited and actively involved in their learning journey.
5. **Parental Involvement**: To support parents and caregivers, the book gives a great foundation to maximize the learning experience.

"**The Exciting World of ABCs and 123s: A Fun Handwriting and Counting Adventure**" is the ideal companion for preschoolers embarking on their educational journey. This book provides a safe and supportive space for young learners to explore handwriting, letter tracing, number recognition, and counting in a fun and rewarding way. Parents, caregivers, and teachers will appreciate the comprehensive and age-appropriate activities that foster early literacy and numeracy skills. So, get ready to embark on a learning adventure filled with tracing joy, numerical excitement, and the delight of mastering new skills. Happy tracing and counting!

Letters
Practice

Letter Tracing

Name: _____

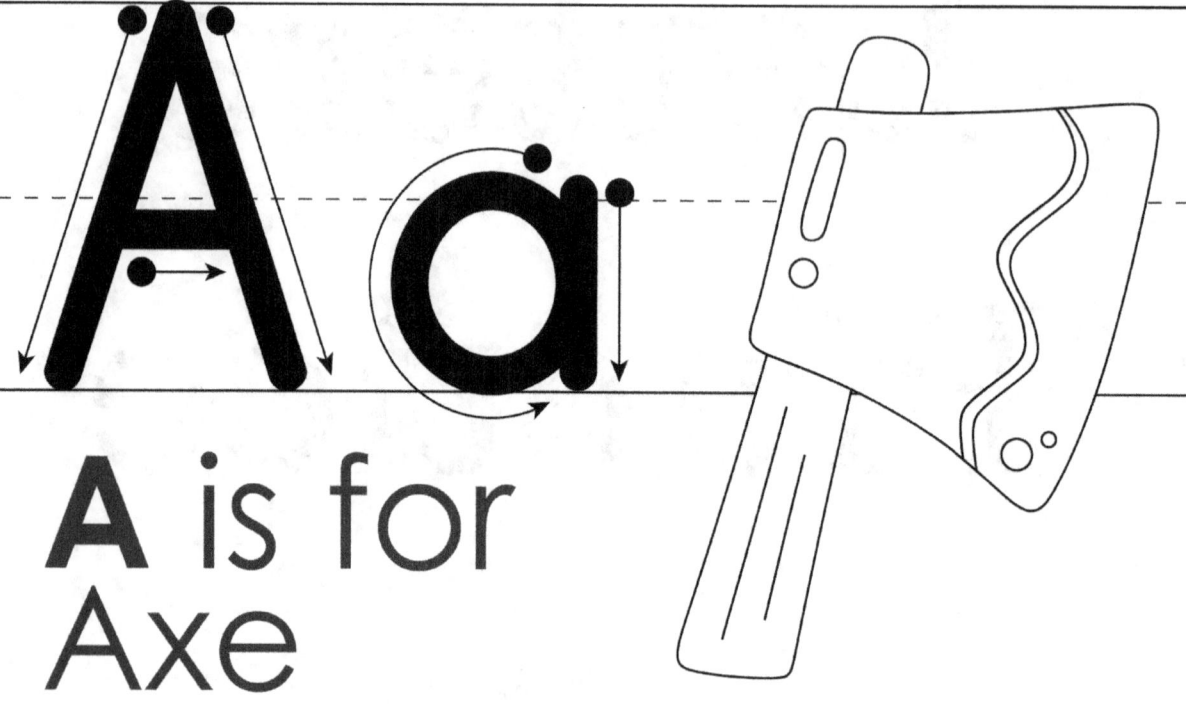

A is for Axe

Trace the letters with a pencil. Then practice writing the letters on the lines

PRACTICE WORKSHEET

Letter Tracing

Name: _____

B B

B is for
Bell

Trace the letters with a pencil. Then practice writing the letters on the lines

B B B B B B B B B B B

B B B B B B B B B B B

b b b b b b b b b b b

b b b b b b b b b b b

PRACTICE WORKSHEET

Letter Tracing

Name: _____

C is for Candy

Trace the letters with a pencil. Then practice writing the letters on the lines

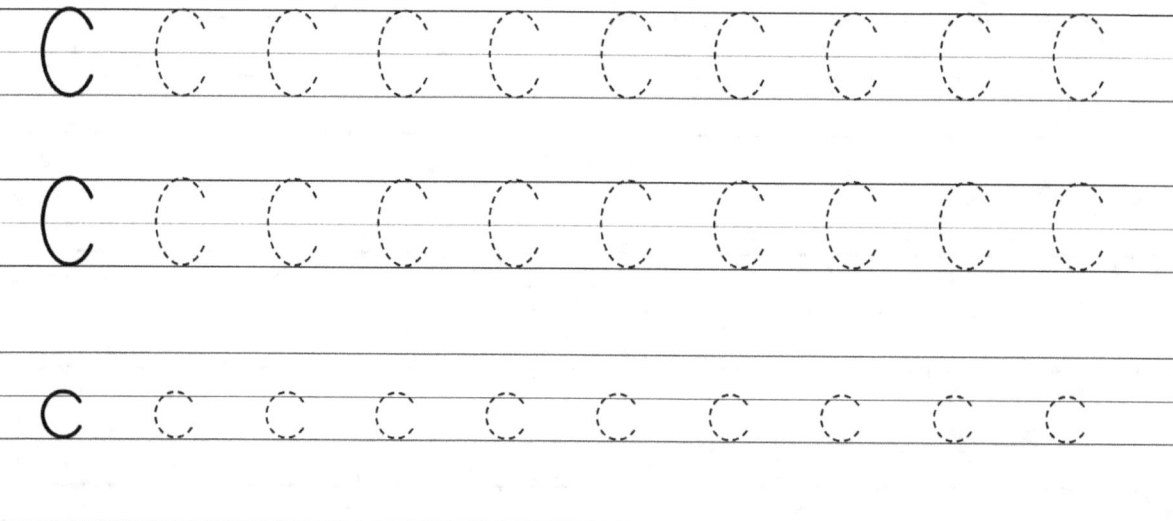

PRACTICE WORKSHEET

Letter Tracing

Name:

D d

D is for
Dice

Trace the letters with a pencil. Then practice writing the letters on the lines

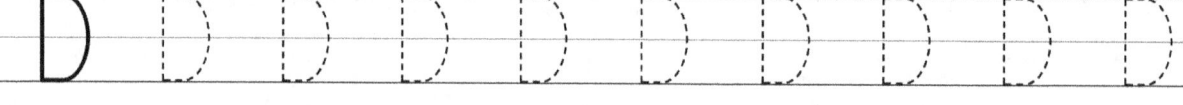

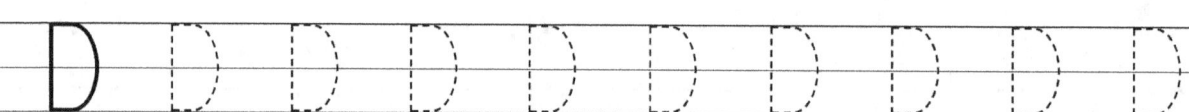

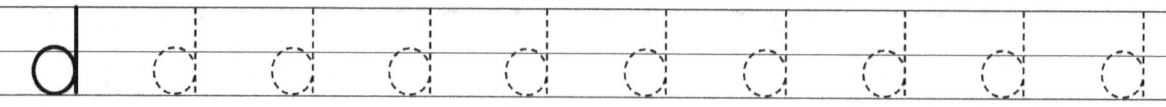

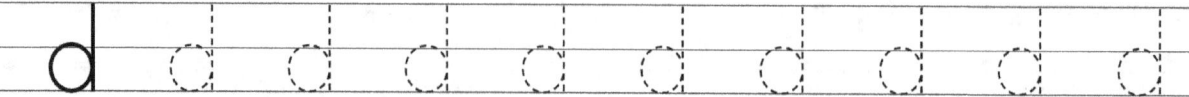

PRACTICE WORKSHEET

Letter Tracing

Name:

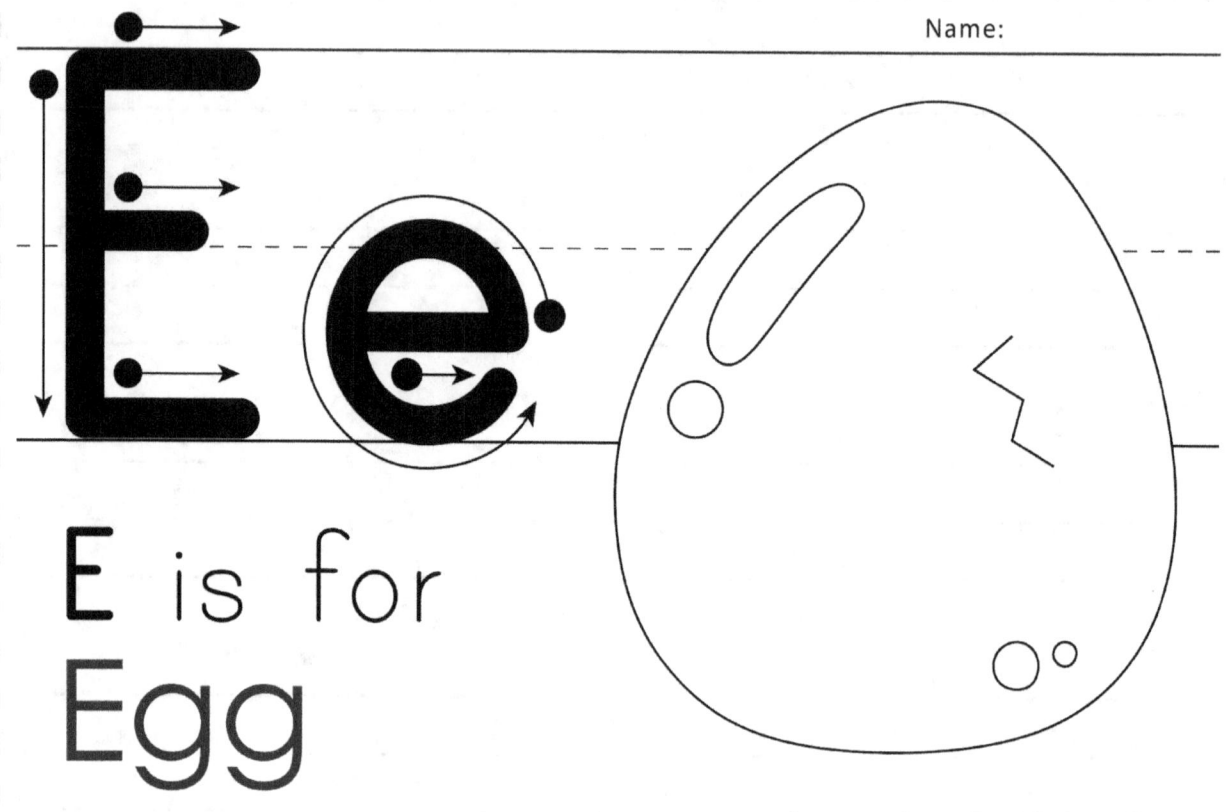

E is for
Egg

Trace the letters with a pencil. Then practice writing the letters on the lines

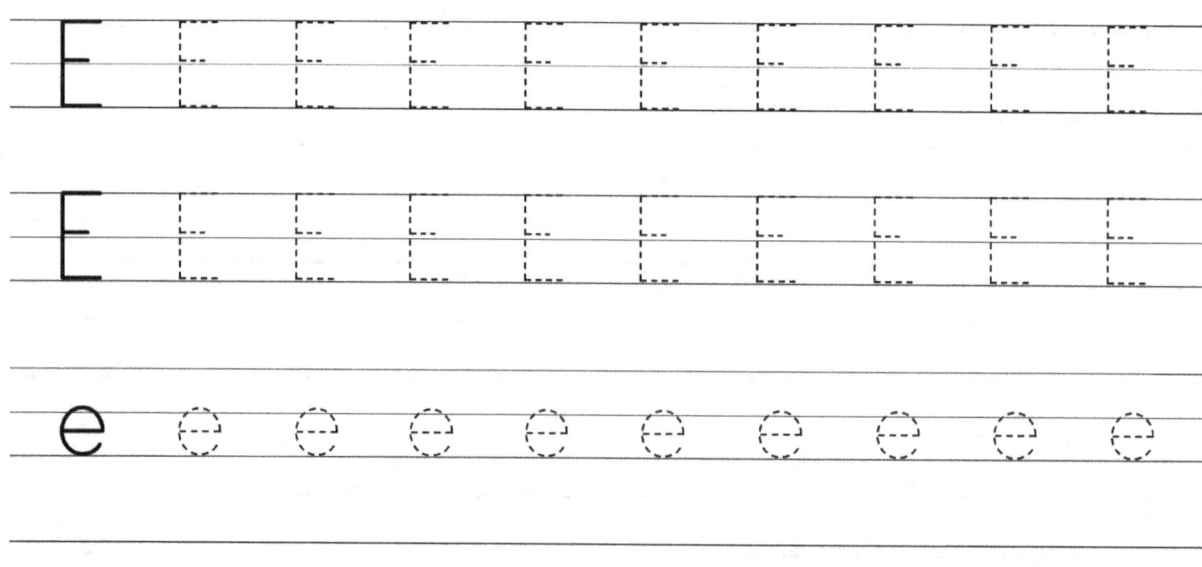

PRACTICE WORKSHEET

Letter Tracing

Name: _____

F is for
Flower

Trace the letters with a pencil. Then practice writing the letters on the lines

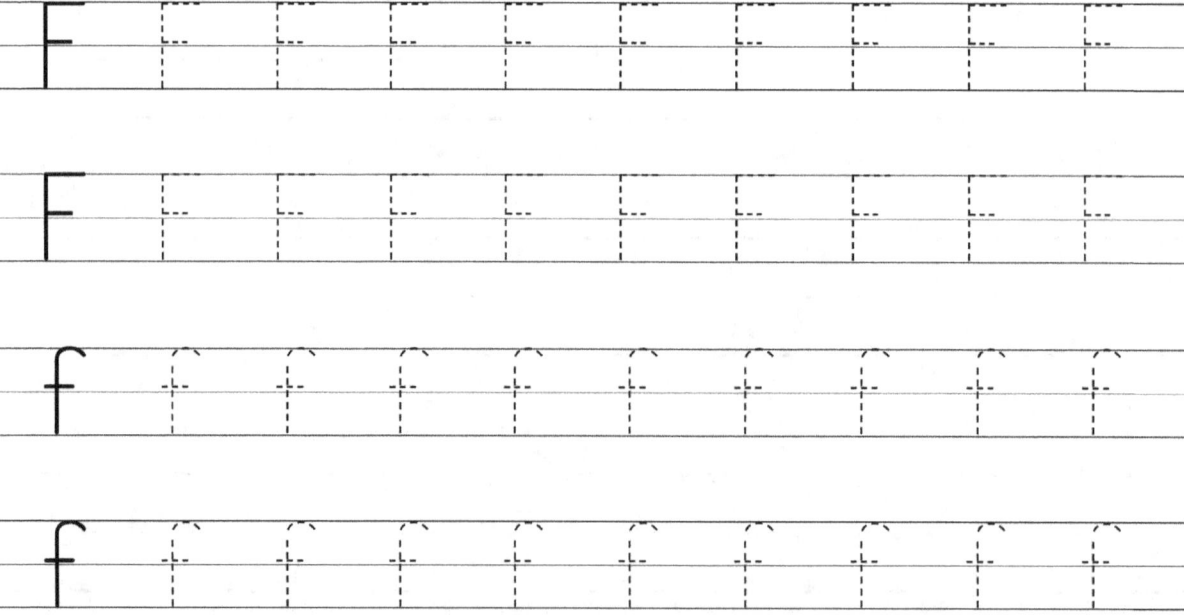

PRACTICE WORKSHEET

Letter Tracing

Name: _____

G g

G is for
Glove

Trace the letters with a pencil. Then practice writing the letters on the lines

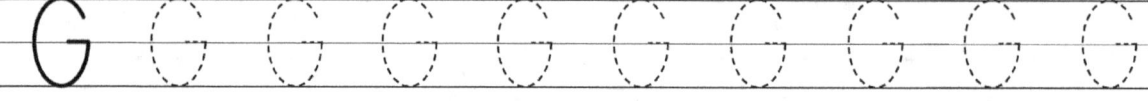

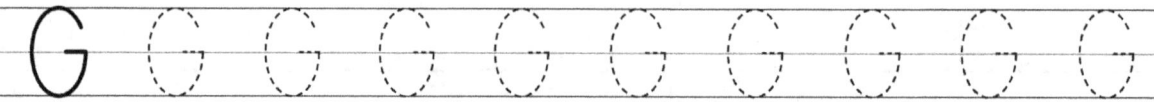

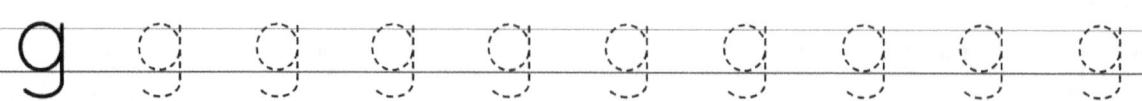

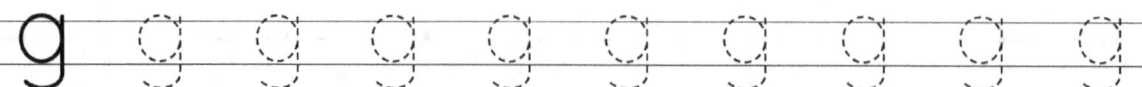

PRACTICE WORKSHEET

Letter Tracing

Name: _____

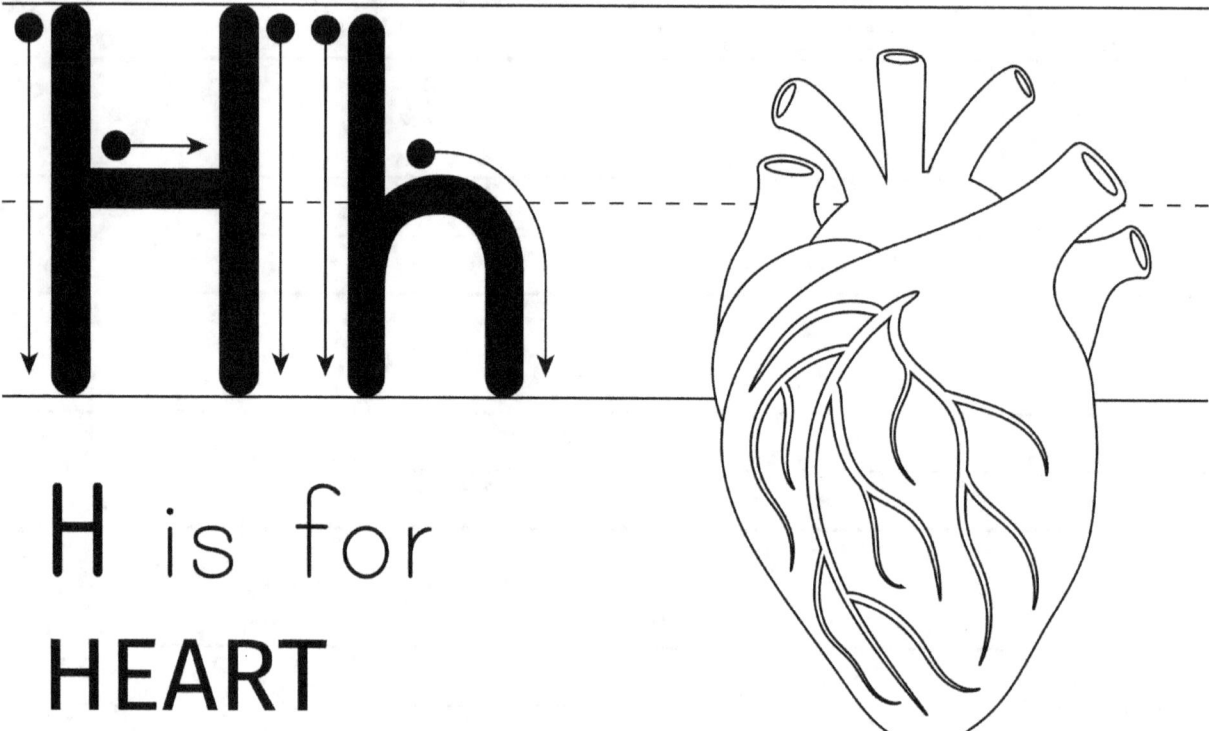

H is for

HEART

Trace the letters with a pencil. Then practice writing the letters on the lines

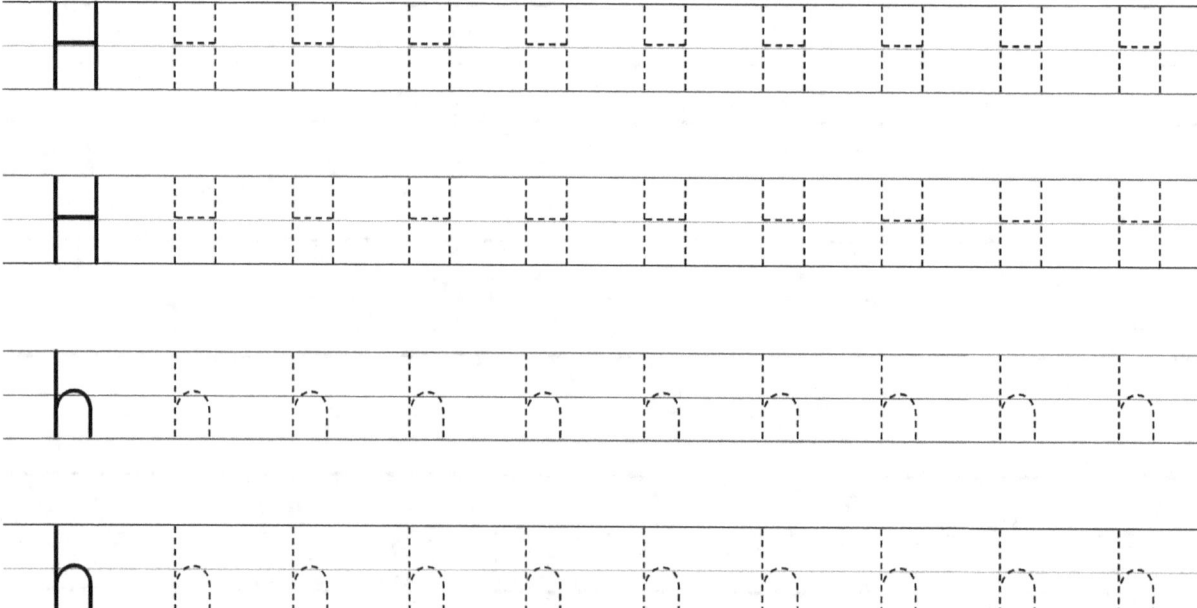

PRACTICE WORKSHEET

Letter Tracing

Name: _____

I is for
Ice cream

Trace the letters with a pencil. Then practice writing the letters on the lines

PRACTICE WORKSHEET

Letter Tracing

Name: _____

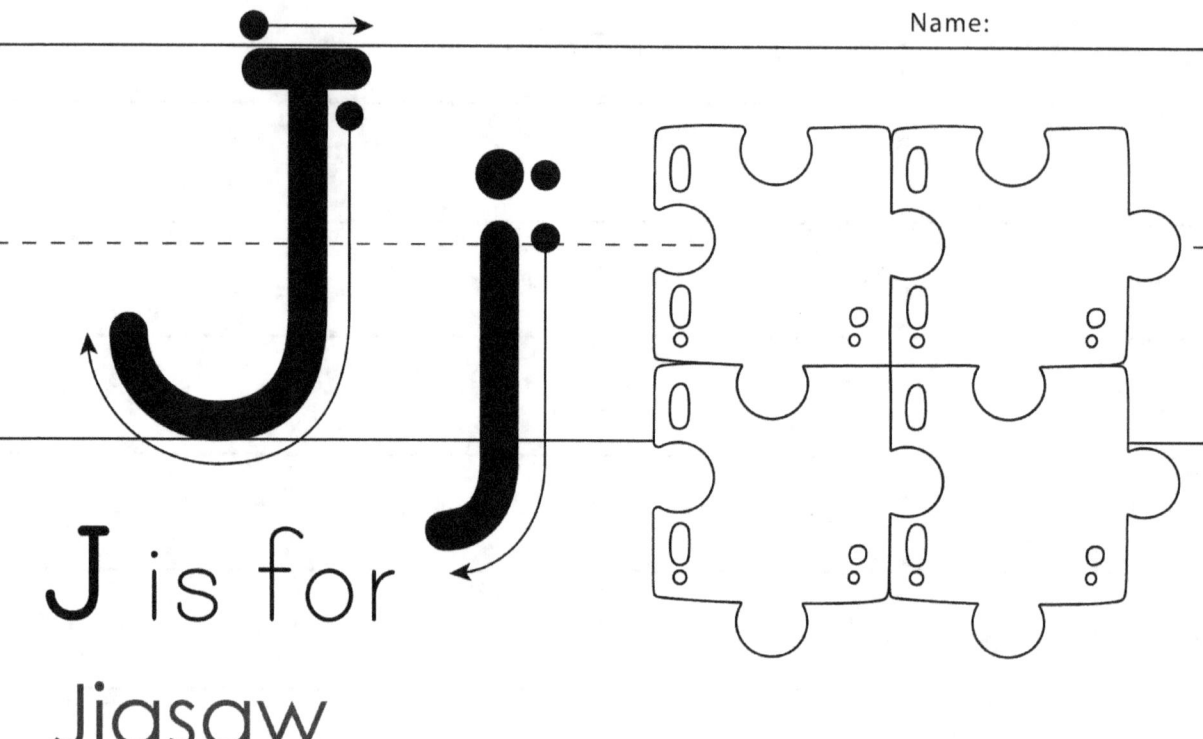

J is for

Jigsaw

Trace the letters with a pencil. Then practice writing the letters on the lines

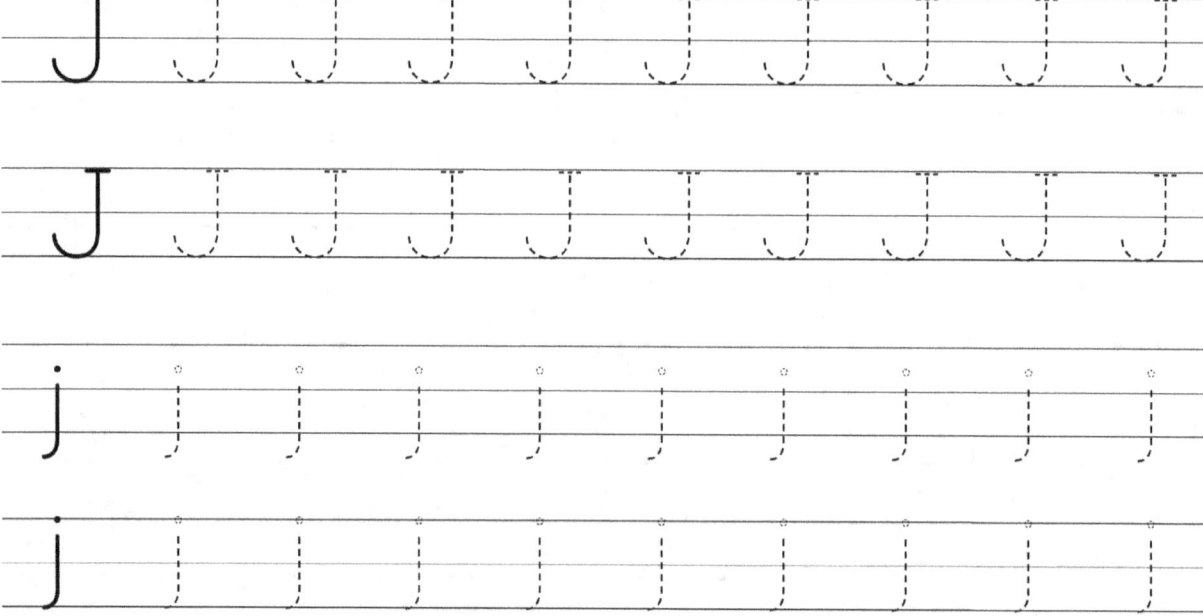

PRACTICE WORKSHEET

Letter Tracing

Name: _____

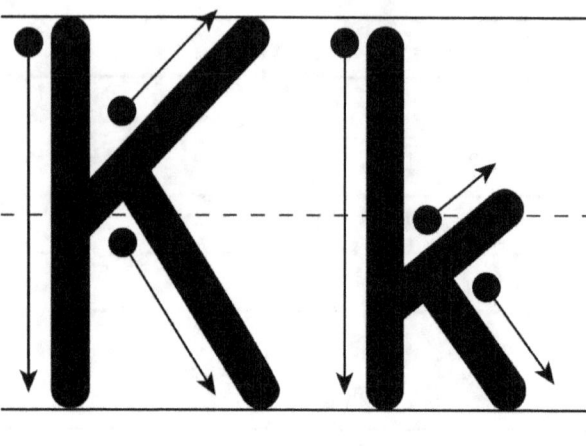

K is for
Ketchup

Trace the letters with a pencil. Then practice writing the letters on the lines

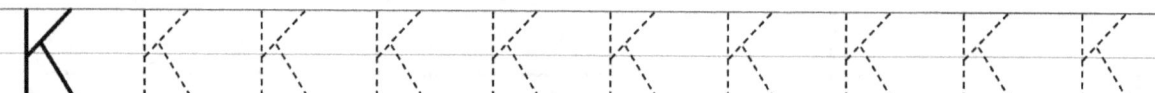

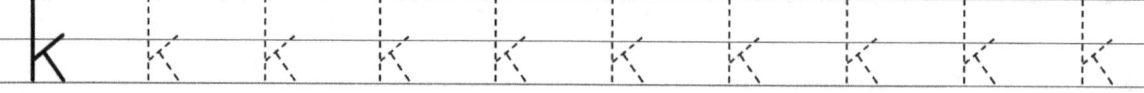

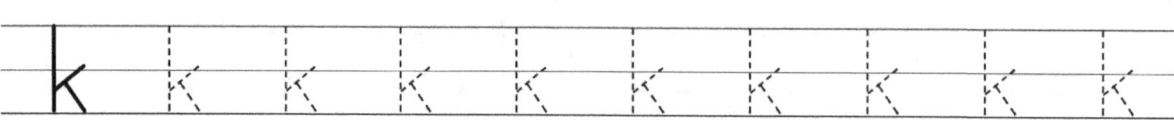

PRACTICE WORKSHEET

Letter Tracing

Name: _____

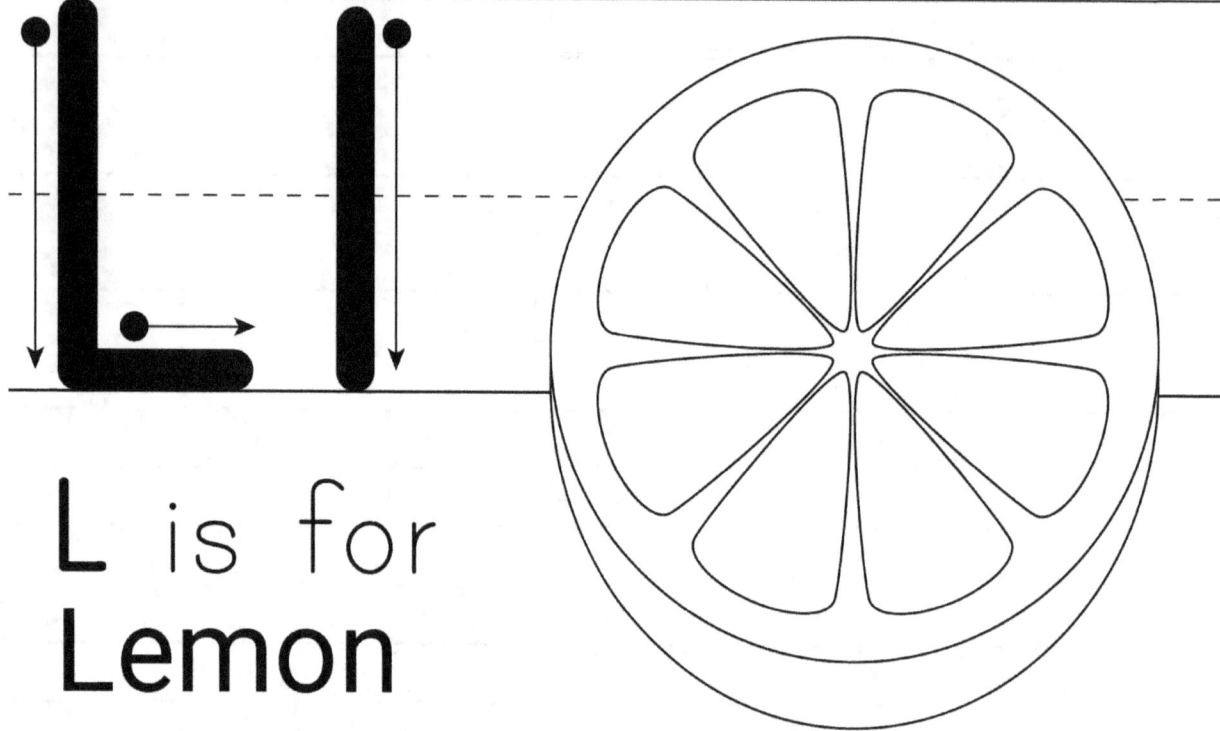

L is for
Lemon

Trace the letters with a pencil. Then practice writing the letters on the lines

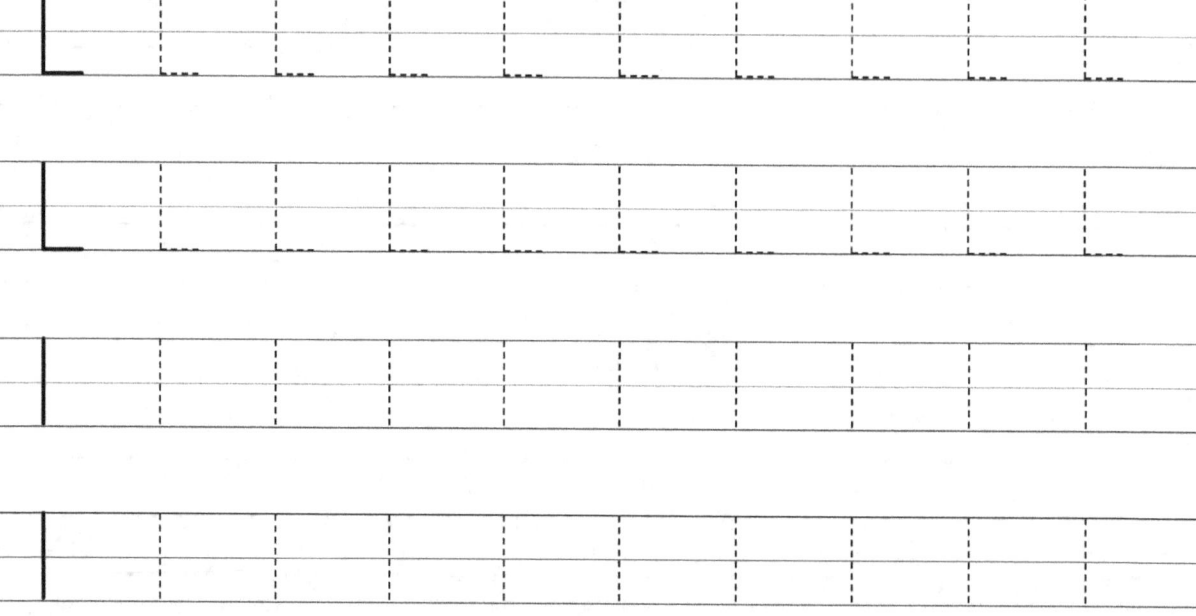

PRACTICE WORKSHEET

Letter Tracing

Name: _____

M is for
Milk

Trace the letters with a pencil. Then practice writing the letters on the lines

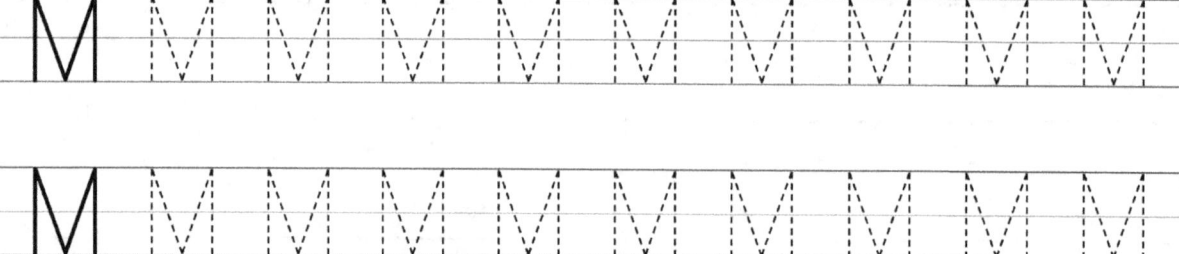

PRACTICE WORKSHEET

Letter Tracing

Name: _____

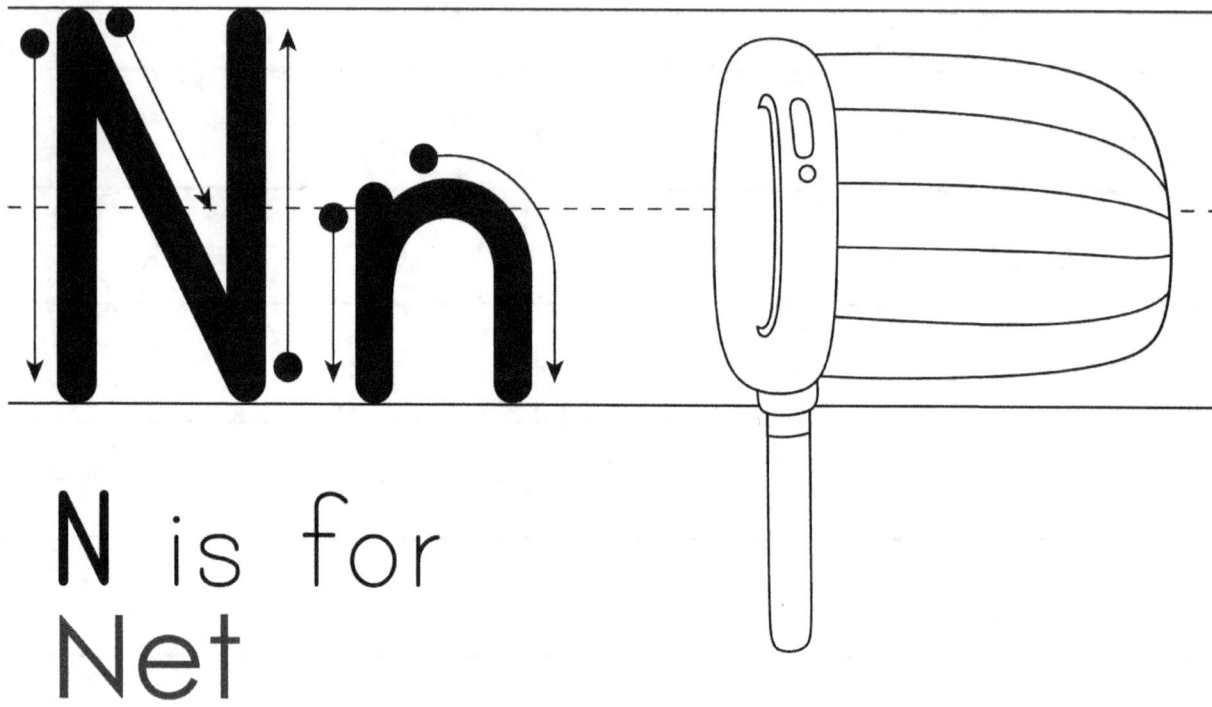

N is for
Net

Trace the letters with a pencil. Then practice writing the letters on the lines

PRACTICE WORKSHEET

Letter Tracing

Name: _____

O is for
Oyster

Trace the letters with a pencil. Then practice writing the letters on the lines

PRACTICE WORKSHEET

Letter Tracing

Name: _____

P is for Pancake

Trace the letters with a pencil. Then practice writing the letters on the lines

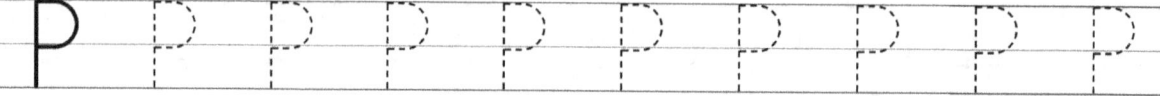

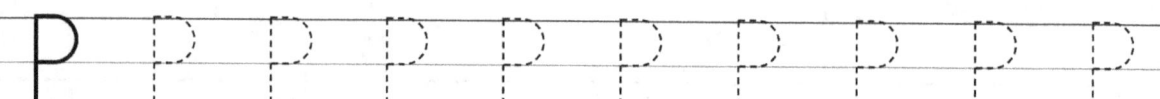

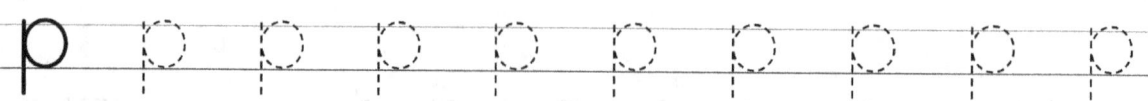

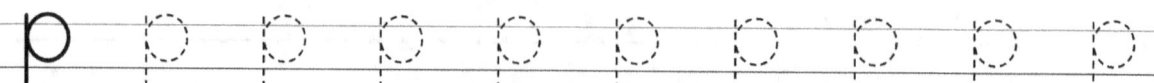

PRACTICE WORKSHEET

Letter Tracing

Name: _____

Q is for
Quiver

Trace the letters with a pencil. Then practice writing the letters on the lines

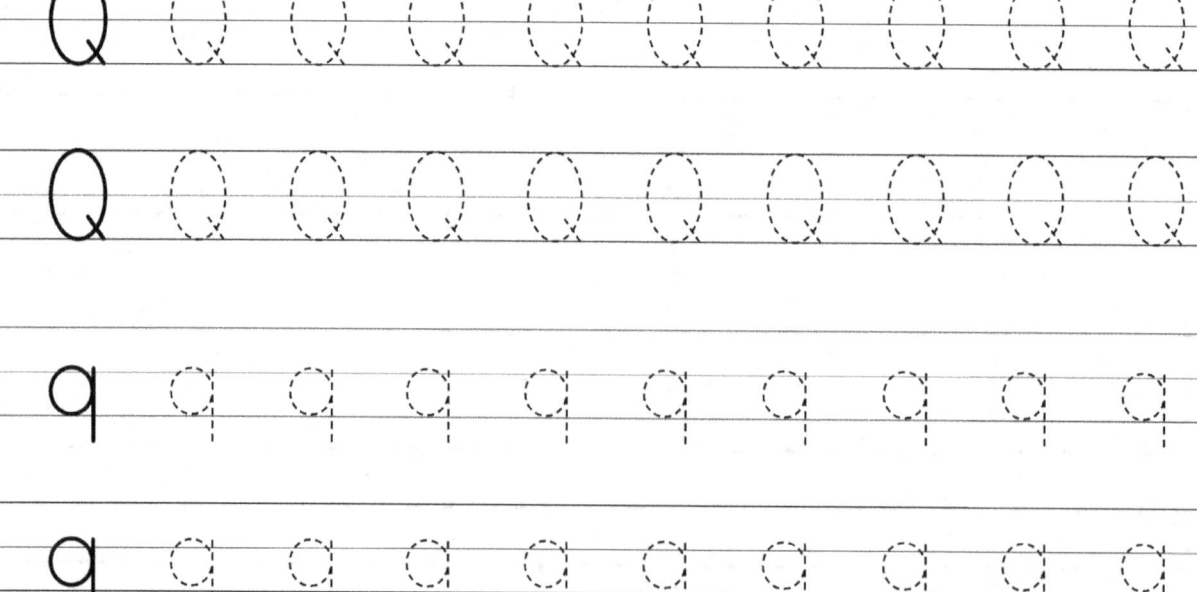

PRACTICE WORKSHEET

Letter Tracing

Name: _____

Rr

R is for
Roller skate

Trace the letters with a pencil. Then practice writing the letters on the lines

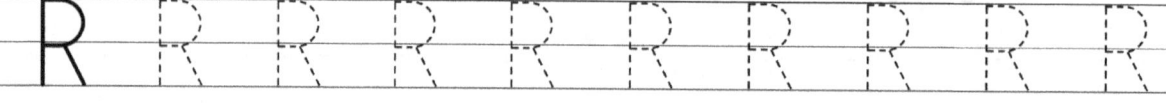

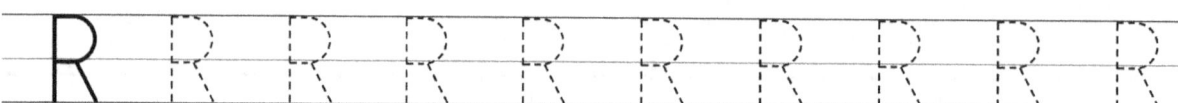

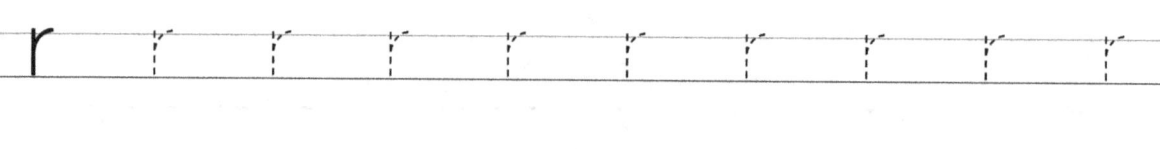

PRACTICE WORKSHEET

Letter Tracing

Name: _____

S s

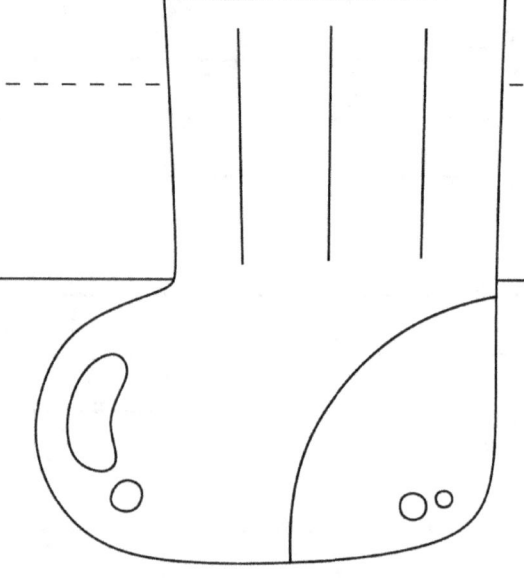

S is for
Sock

Trace the letters with a pencil. Then practice writing the letters on the lines

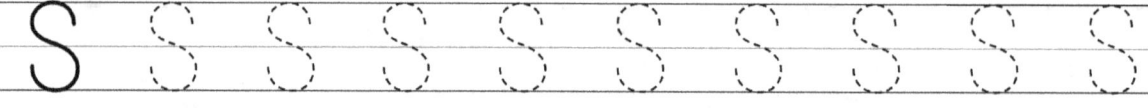

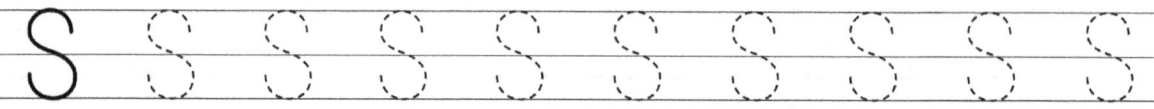

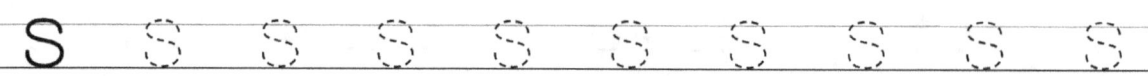

PRACTICE WORKSHEET

Letter Tracing

Name: _____

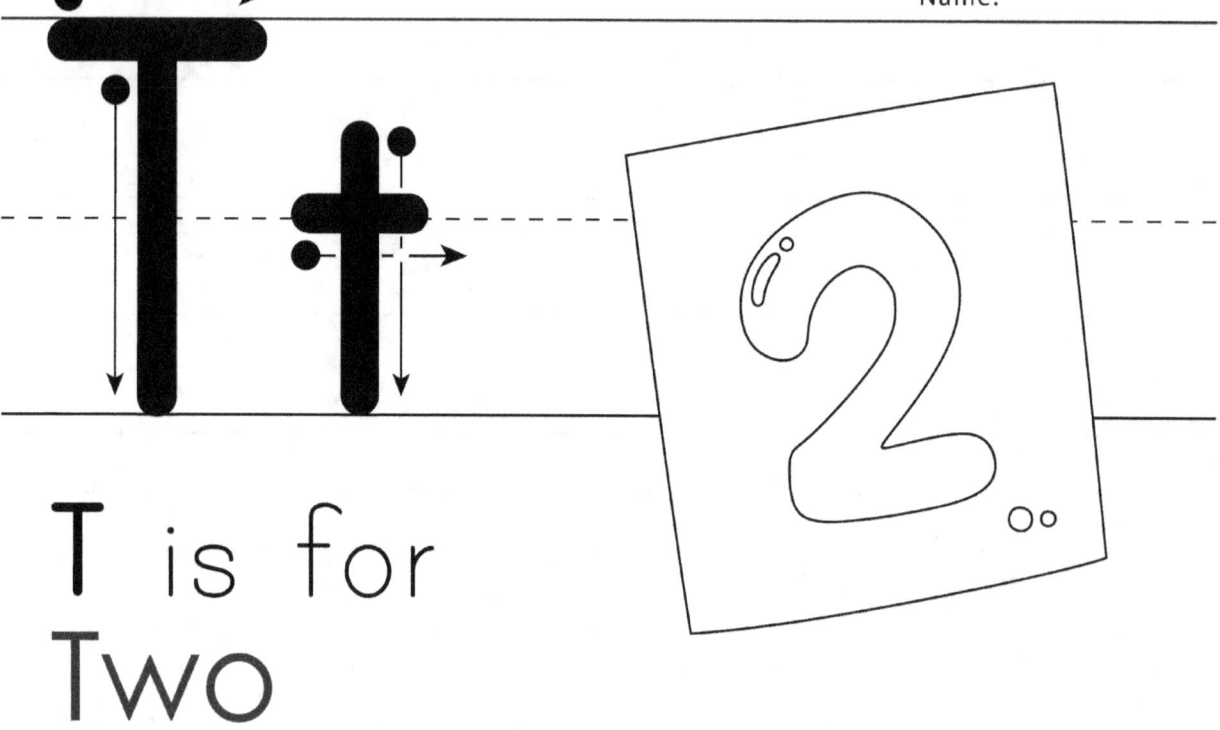

T is for
Two

Trace the letters with a pencil. Then practice writing the letters on the lines

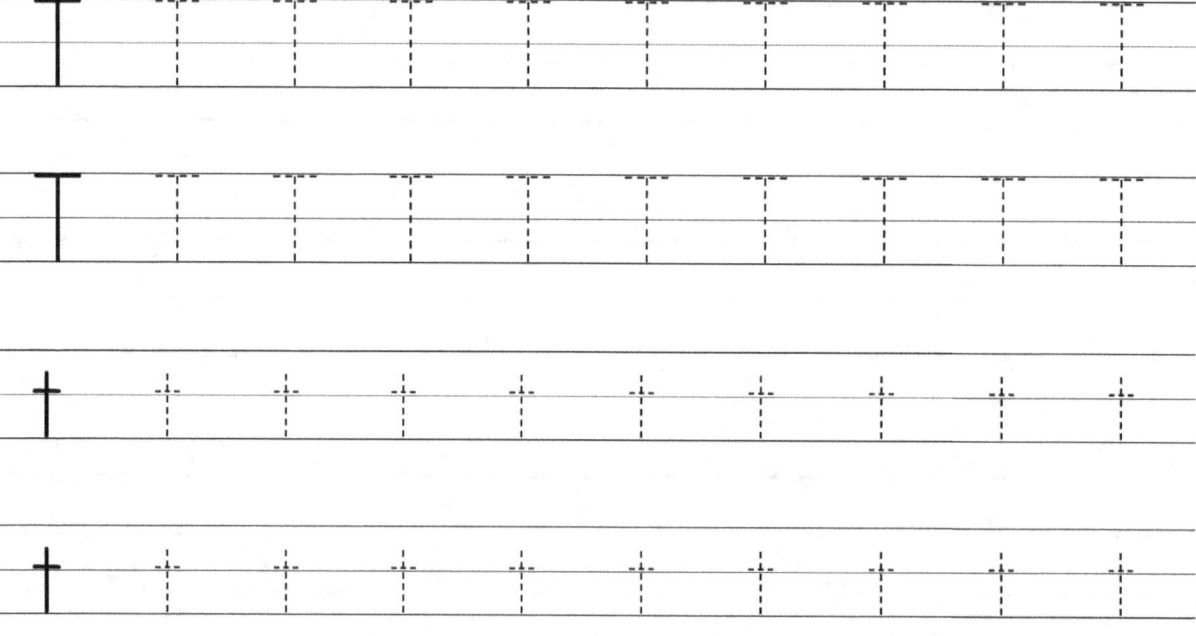

PRACTICE WORKSHEET

Letter Tracing

Name: _____

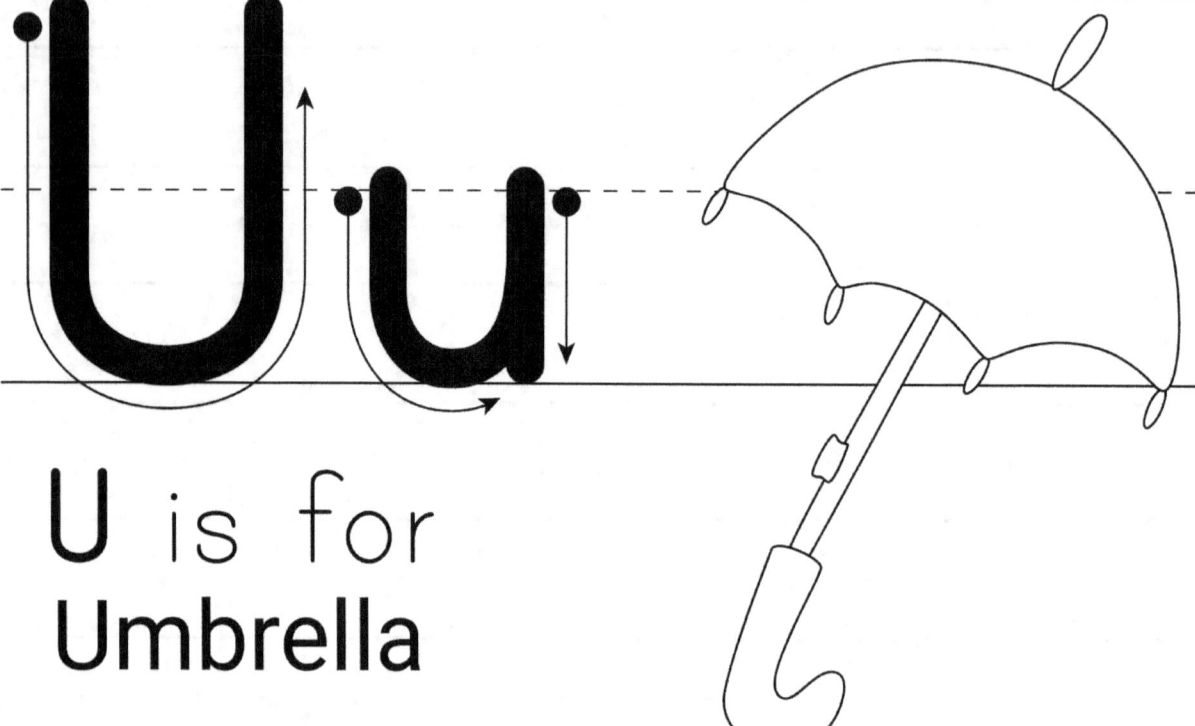

U is for
Umbrella

Trace the letters with a pencil. Then practice writing the letters on the lines

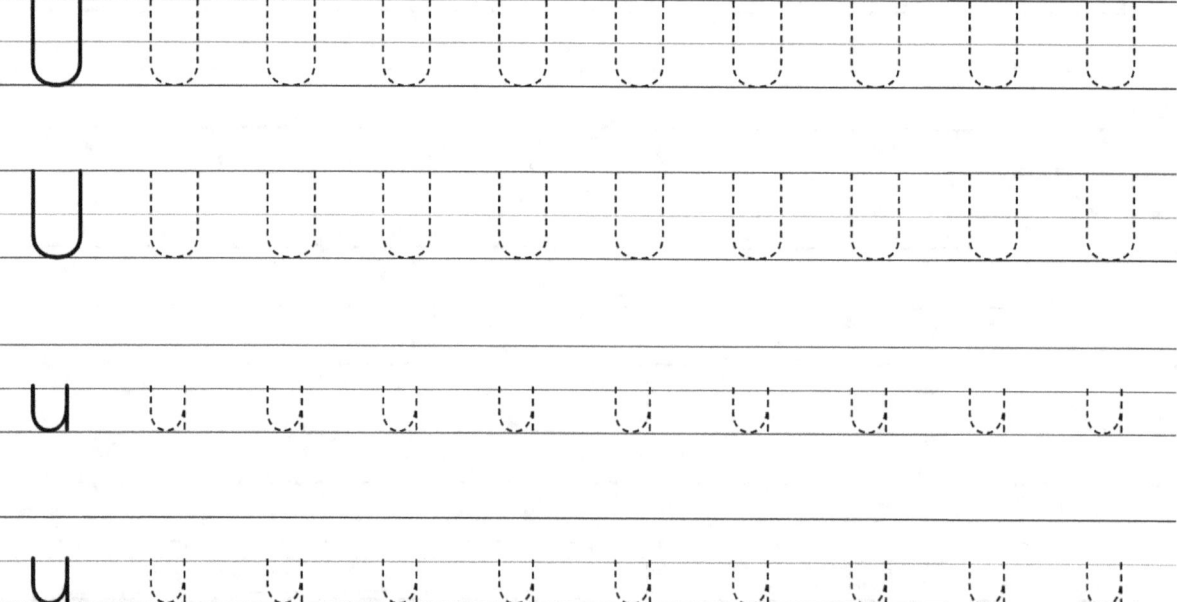

PRACTICE WORKSHEET

Letter Tracing

Name: _____

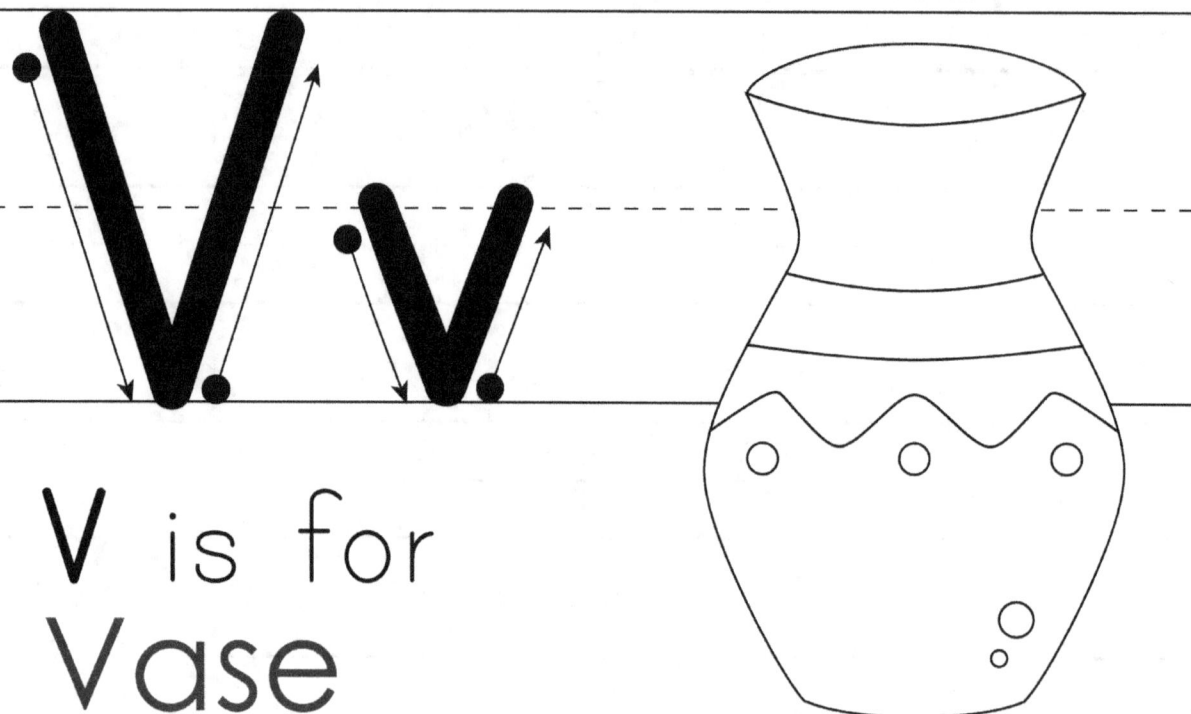

V is for
Vase

Trace the letters with a pencil. Then practice writing the letters on the lines

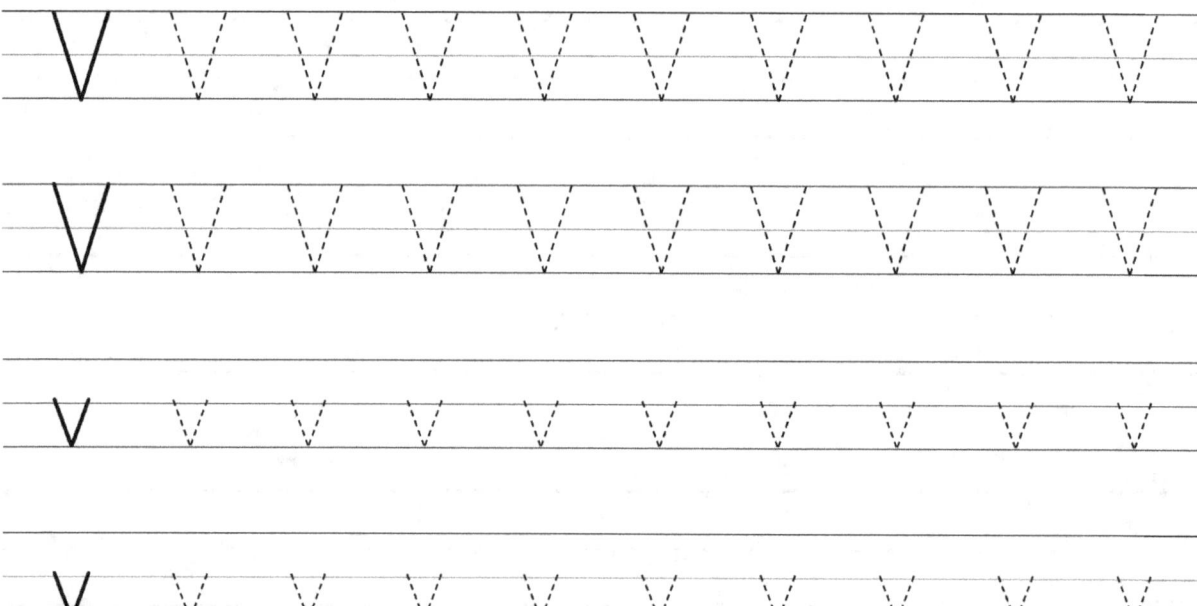

PRACTICE WORKSHEET

Letter Tracing

Name: _____

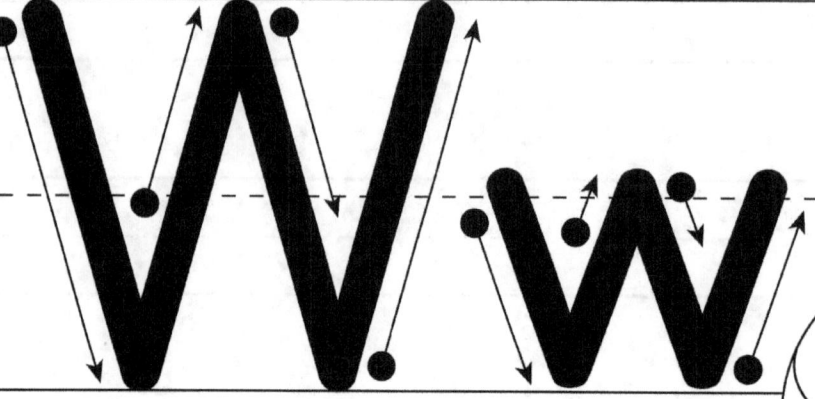

W is for
Wind wheel

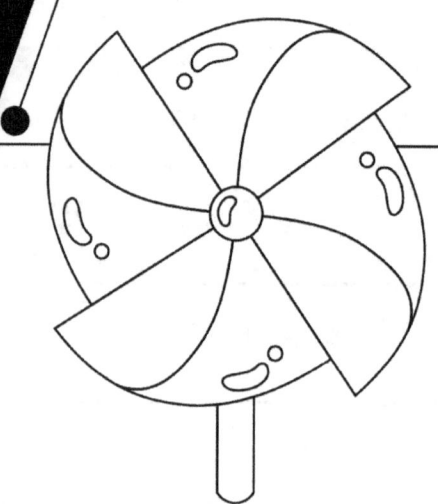

Trace the letters with a pencil. Then practice writing the letters on the lines

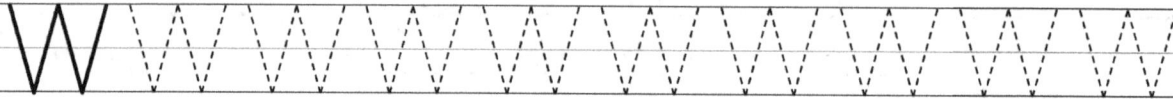

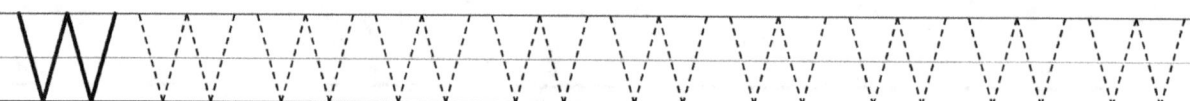

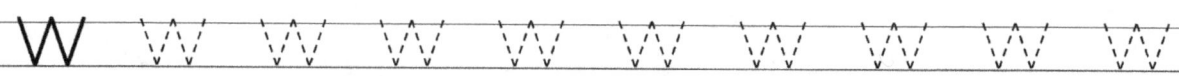

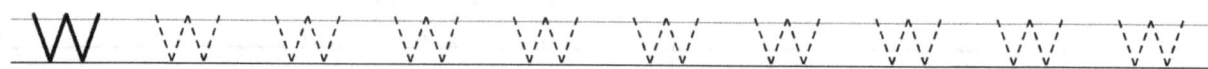

PRACTICE WORKSHEET

Letter Tracing

Name:

X ON THE DOOR

Trace the letters with a pencil. Then practice writing the letters on the lines

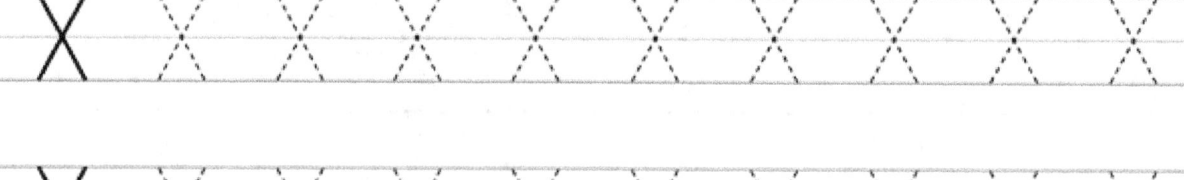

PRACTICE WORKSHEET

Letter Tracing

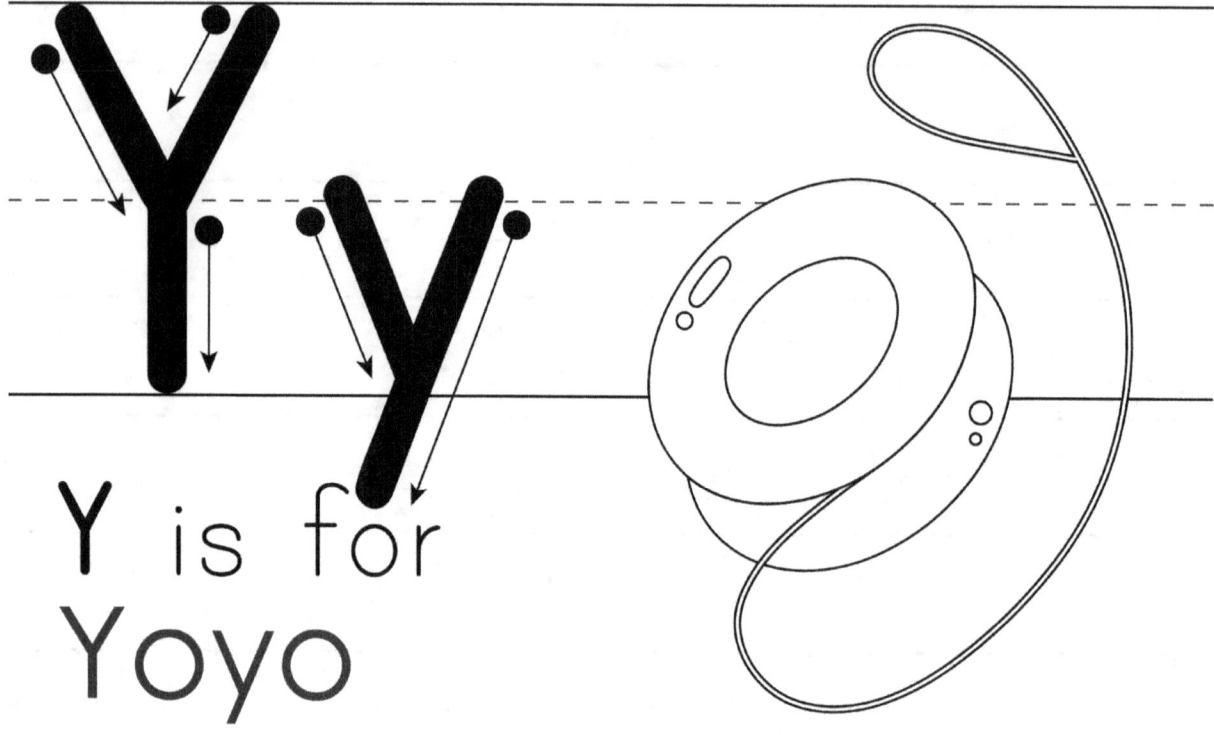

Y is for
Yoyo

Trace the letters with a pencil. Then practice writing the letters on the lines

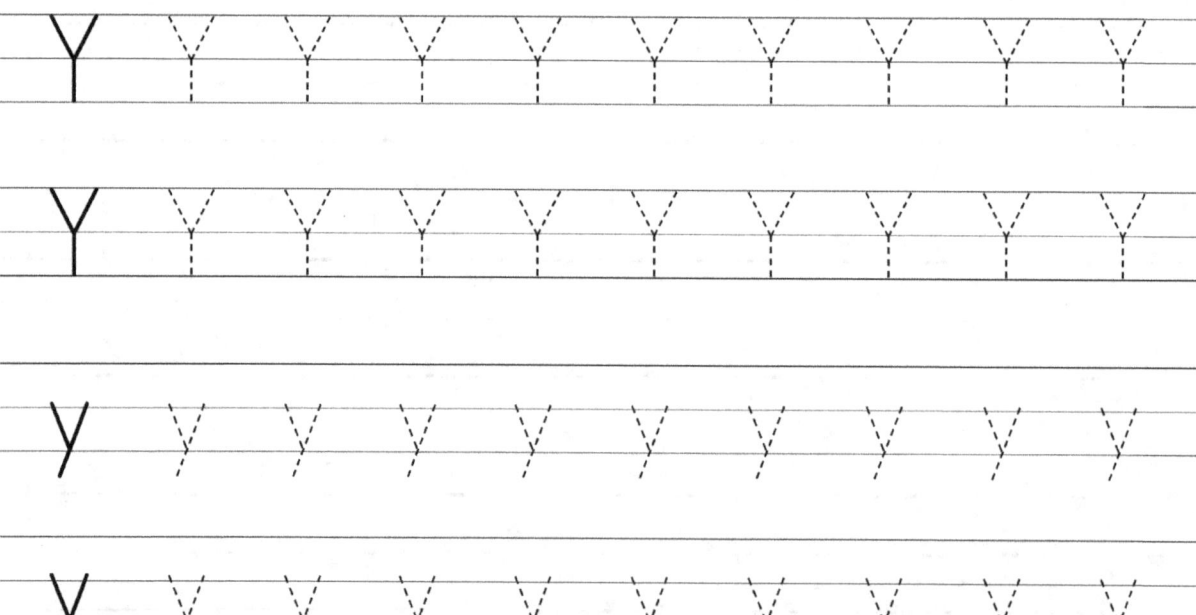

PRACTICE WORKSHEET

Letter Tracing

Name: _____

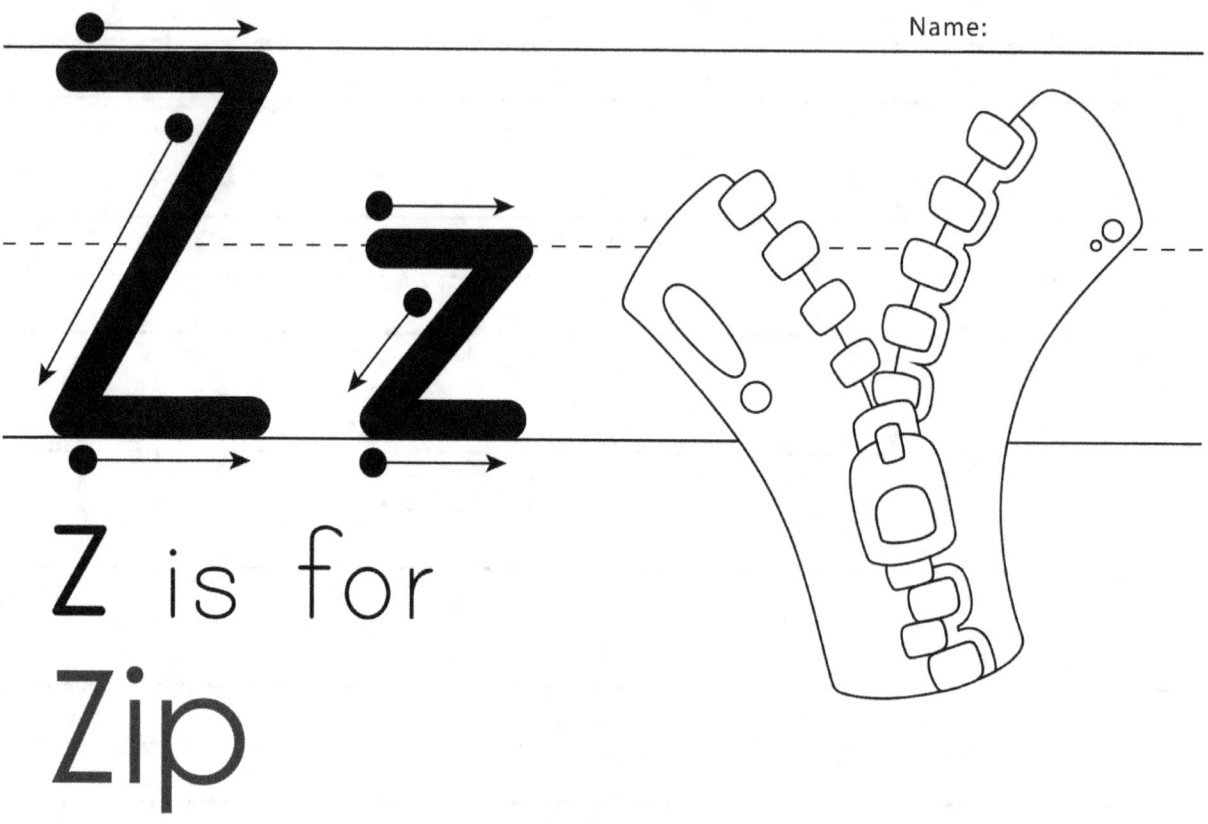

Z is for Zip

Trace the letters with a pencil. Then practice writing the letters on the lines

PRACTICE WORKSHEET

Skills
Page

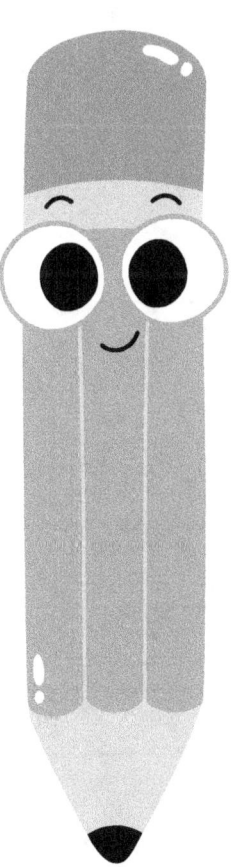

Tracing Letters

Directions: Trace all the upper and lower case letters.

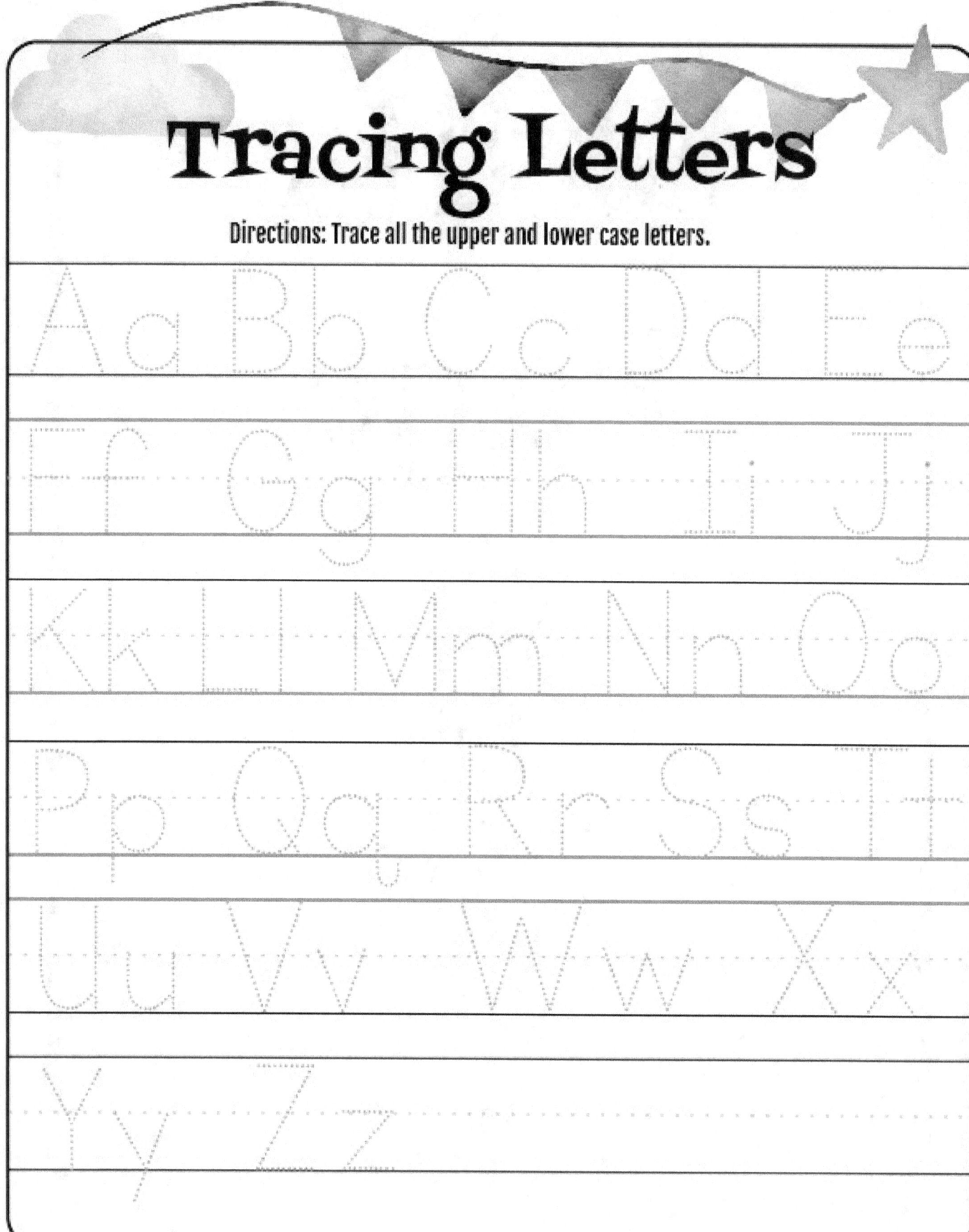

PRACTICE WRITING YOUR LOWERCASE LETTERS.

a _____ b _____ c _____ d _____

e _____ f _____ g _____ h _____

i _____ j _____ k _____ l _____

m _____   n _____

o _____ p _____ q _____ r _____

s _____ t _____ u _____ v _____

w _____ x _____ y _____ z _____

ABC ORDER MAZE

ORDER MAZE

START →	A M	A	P	T
L P	C	B A	R	D B
F E	D	B A	P	T C
G H	A L	M	R S	A
U I	J	K N	O T	O
M Z	X R	Q	P C	V
L E	T S	R	B N	E
A V	U N	S	R S	D
Y W	X Y	Z	→	END

Fill in the missing letters:

A __ C D __ F G H __ J K __ M

N O __ Q R __ T U __ W X __ Z

Numbers Practice

Let's Count!

1 2 3 4 5

COUNT, TRACE AND COLOR

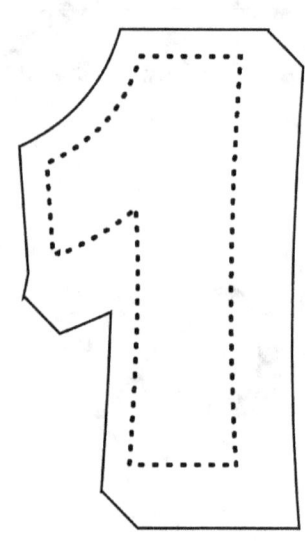 

One apple

1

one one one one

PRACTICE WORKSHEET

COUNT, TRACE AND COLOR

Two beetroot

2 2 2 2 2 2 2

2 2 2 2 2 2 2

two two two two

PRACTICE WORKSHEET

COUNT, TRACE AND COLOR

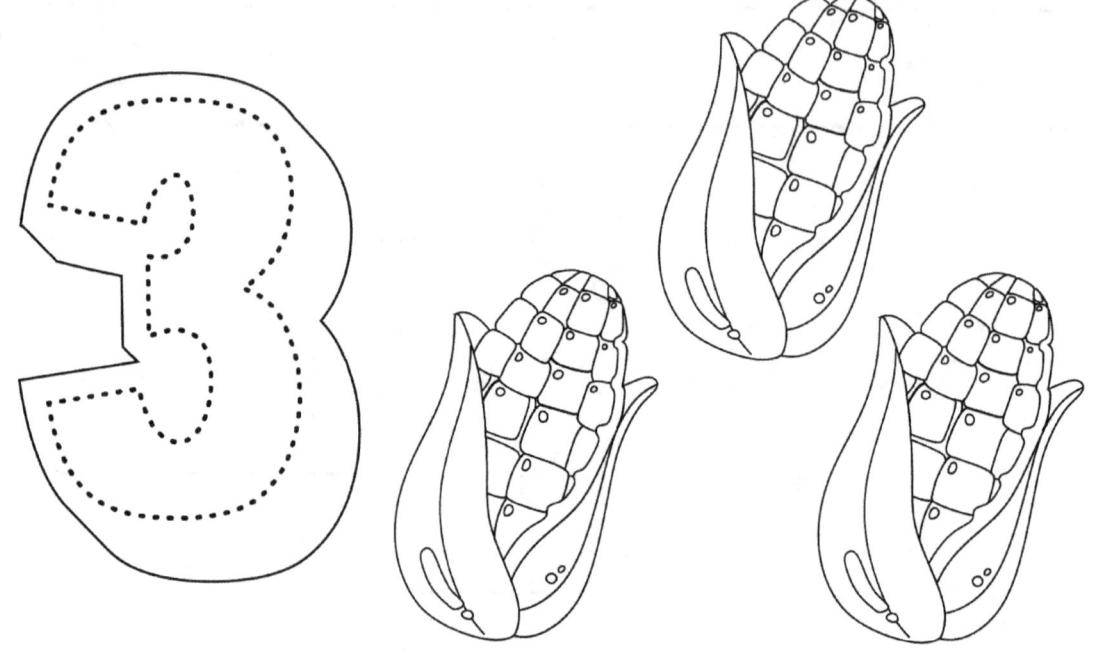

Three corns

3 3 3 3 3 3 3

3 3 3 3 3 3 3

three three three

PRACTICE WORKSHEET

COUNT, TRACE AND COLOR

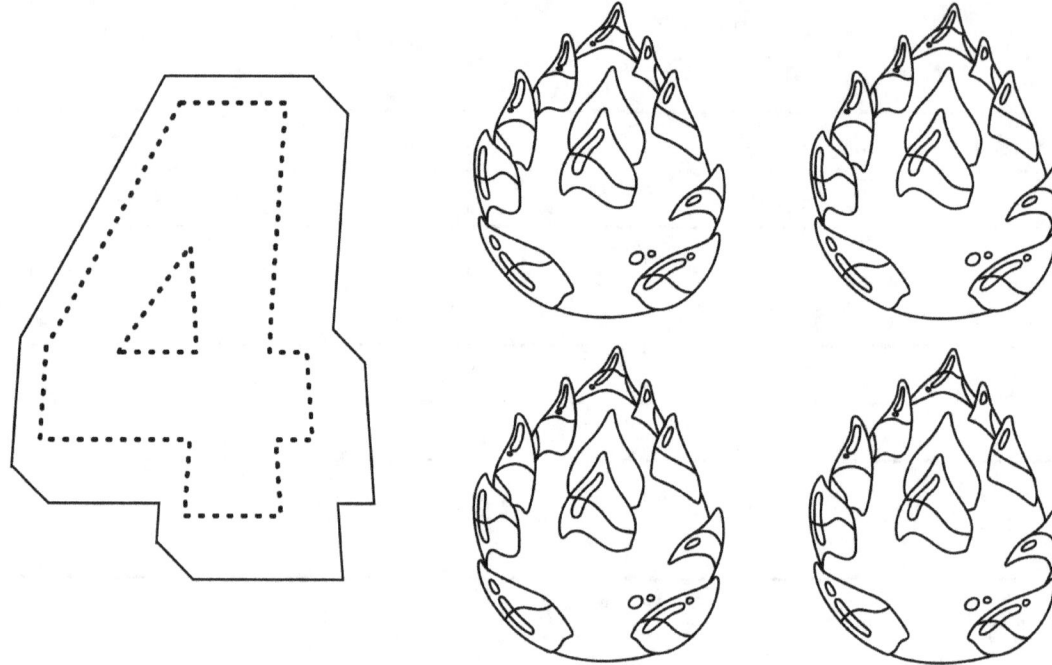

Four dragon fruits

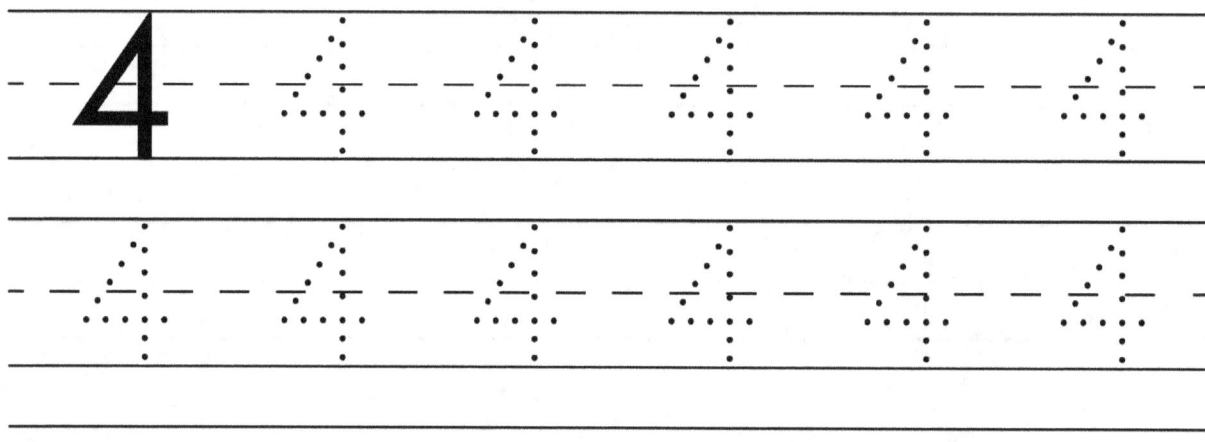

PRACTICE WORKSHEET

COUNT, TRACE AND COLOR

Five eggplants

5 5 5 5 5 5 5

5 5 5 5 5 5 5

five five five five

PRACTICE WORKSHEET

COUNT, TRACE AND COLOR

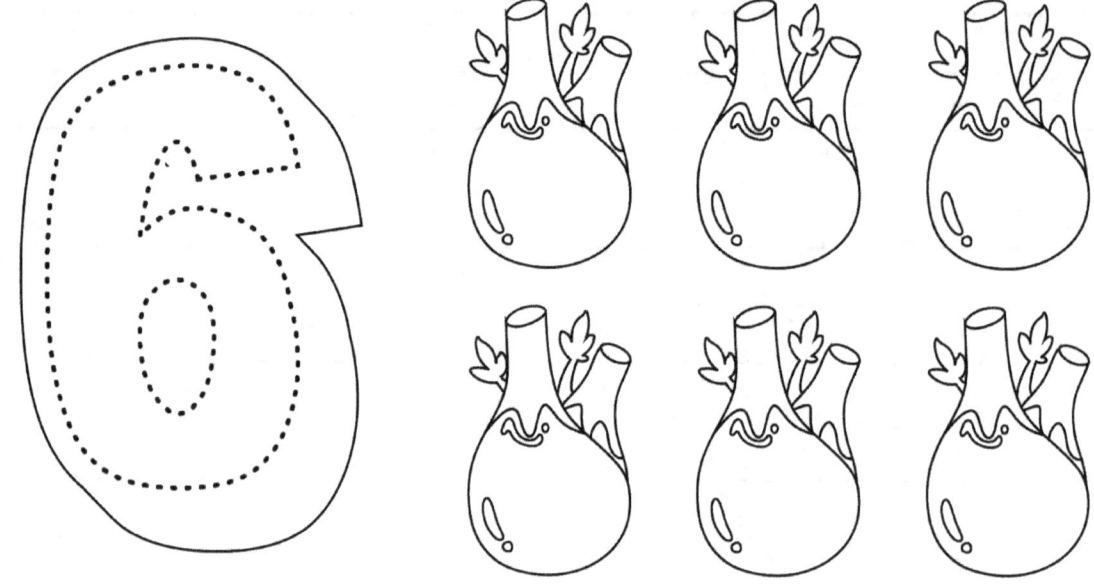

Six fennels

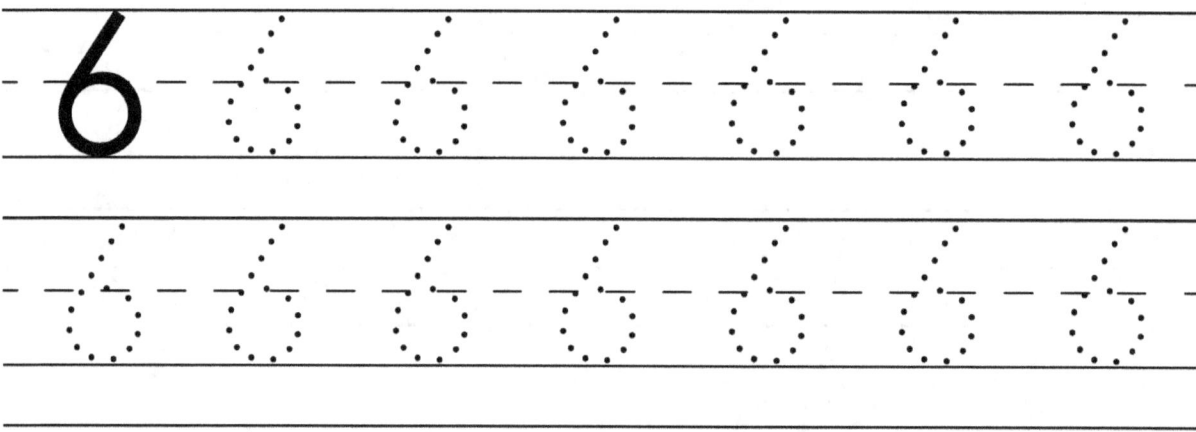

PRACTICE WORKSHEET

COUNT, TRACE AND COLOR

Seven grapes

7

seven

PRACTICE WORKSHEET

COUNT, TRACE AND COLOR

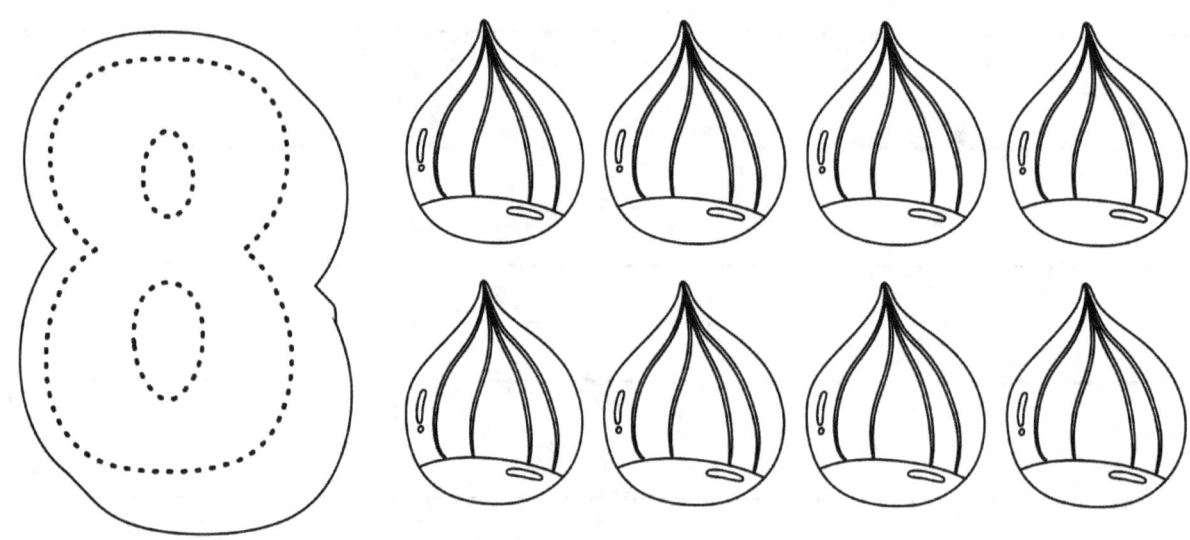

Eight hazelnuts

8

eight

PRACTICE WORKSHEET

COUNT, TRACE AND COLOR

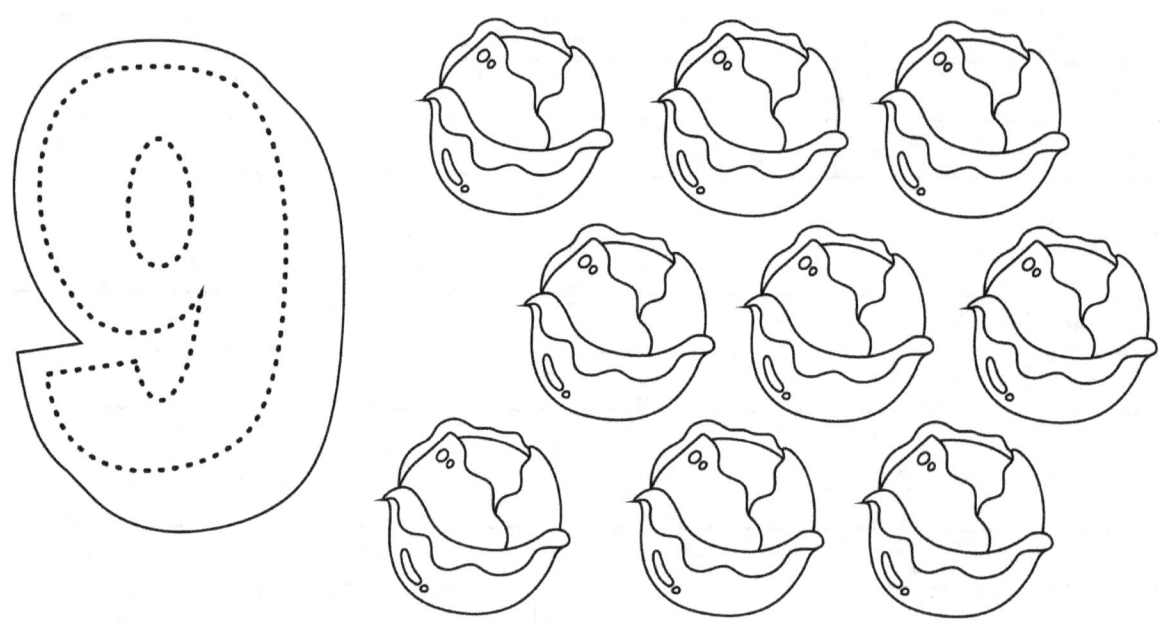

Nine iceberg lettuces

9 9 9 9 9 9 9 9

9 9 9 9 9 9 9

nine nine nine

PRACTICE WORKSHEET

COUNT, TRACE AND COLOR

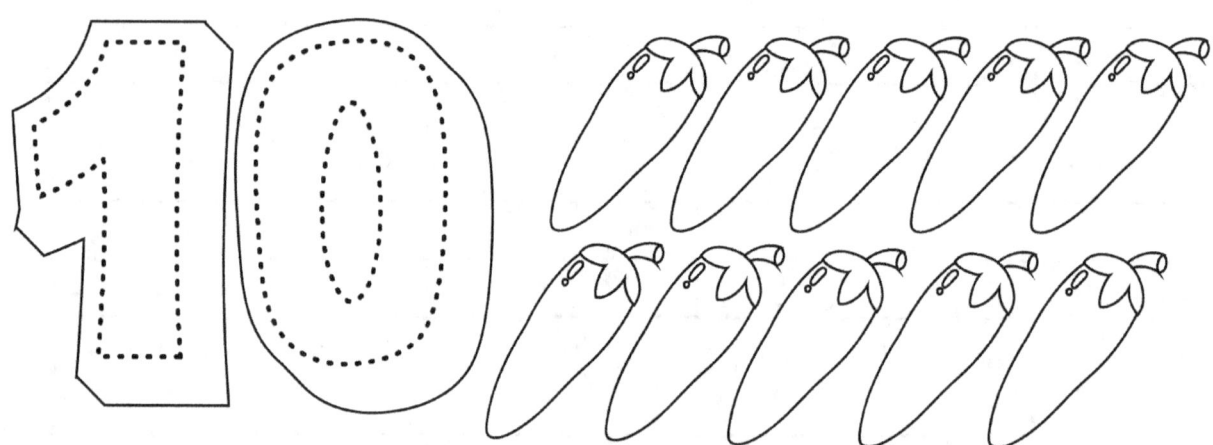

Ten jalapenos

10

ten

PRACTICE WORKSHEET

COUNT, TRACE AND COLOR

Eleven kales

11

eleven

PRACTICE WORKSHEET

COUNT, TRACE AND COLOR

Twelve lemons

12   12 12 12 12

12 12 12 12 12

twelve   twelve

PRACTICE WORKSHEET

COUNT, TRACE AND COLOR

Thirteen mangos

13

thirteen

PRACTICE WORKSHEET

COUNT, TRACE AND COLOR

Fourteen nuts

14   14   14   14   14

14   14   14   14   14

fourteen   fourteen

PRACTICE WORKSHEET

COUNT, TRACE AND COLOR

Fifteen oranges

15 15 15 15 15

15 15 15 15 15

fifteen fifteen

PRACTICE WORKSHEET

COUNT, TRACE AND COLOR

Sixteen peaches

16

sixteen

PRACTICE WORKSHEET

COUNT, TRACE AND COLOR

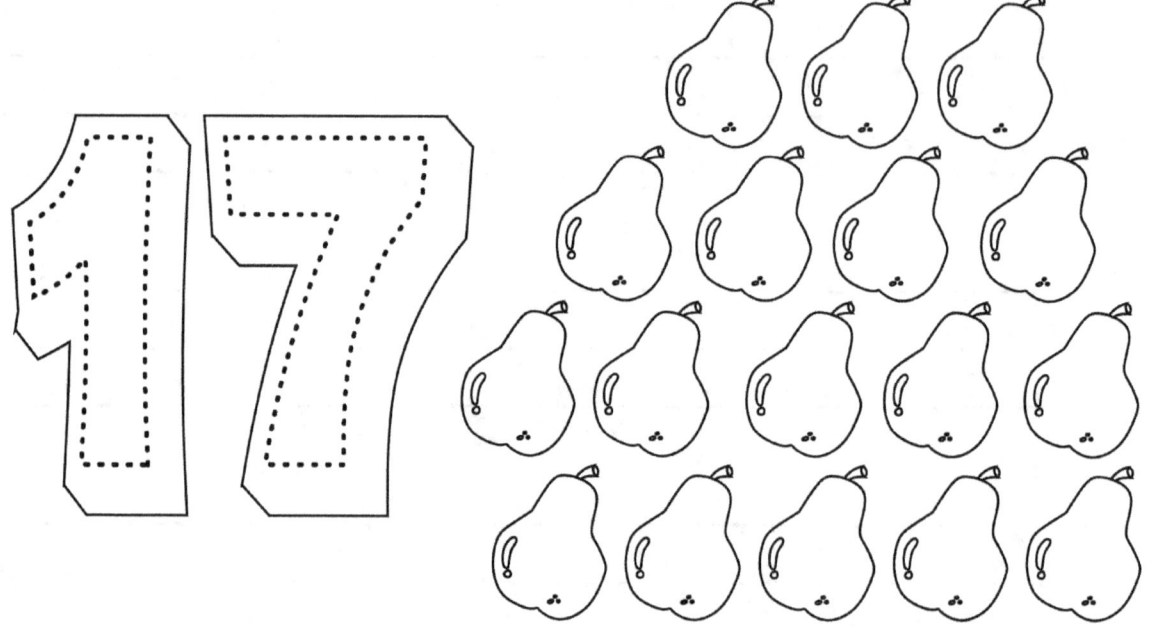

Seventeen quinces

17

PRACTICE WORKSHEET

COUNT, TRACE AND COLOR

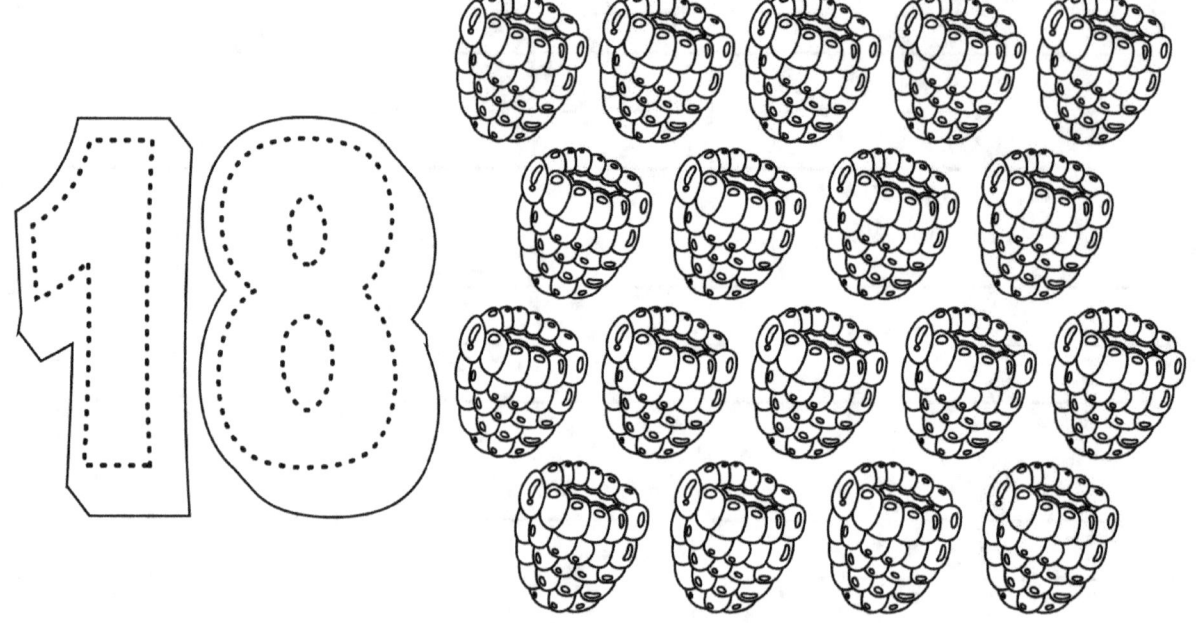

Eighteen raspberries

18 18 18 18 18 18

18 18 18 18 18 18

eighteen

PRACTICE WORKSHEET

COUNT, TRACE AND COLOR

Nineteen strawberries

19

PRACTICE WORKSHEET

COUNT, TRACE AND COLOR

Twenty tomatos

20 20 20 20 20

20 20 20 20

twenty twenty

PRACTICE WORKSHEET

Learning Shapes

Write the names of the shapes

triangle

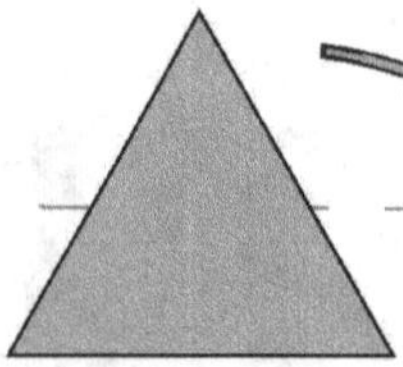

circle

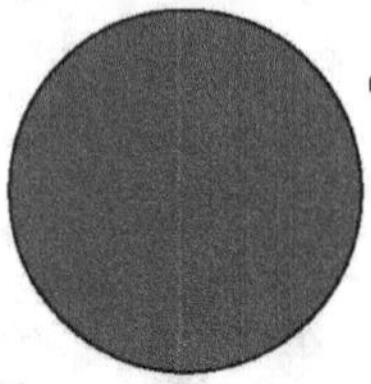

square

oval

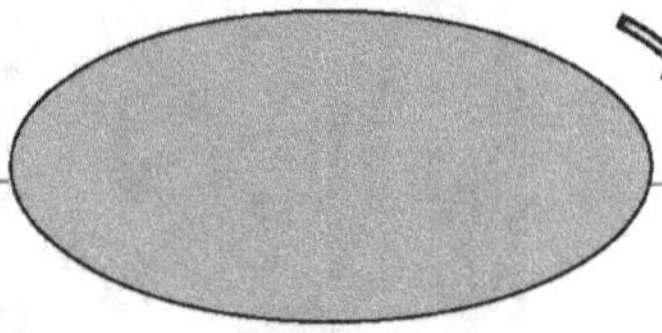

star

rectangle

NAME:...................

Trace and color the squares

NAME:.........................

Trace and color the triangles

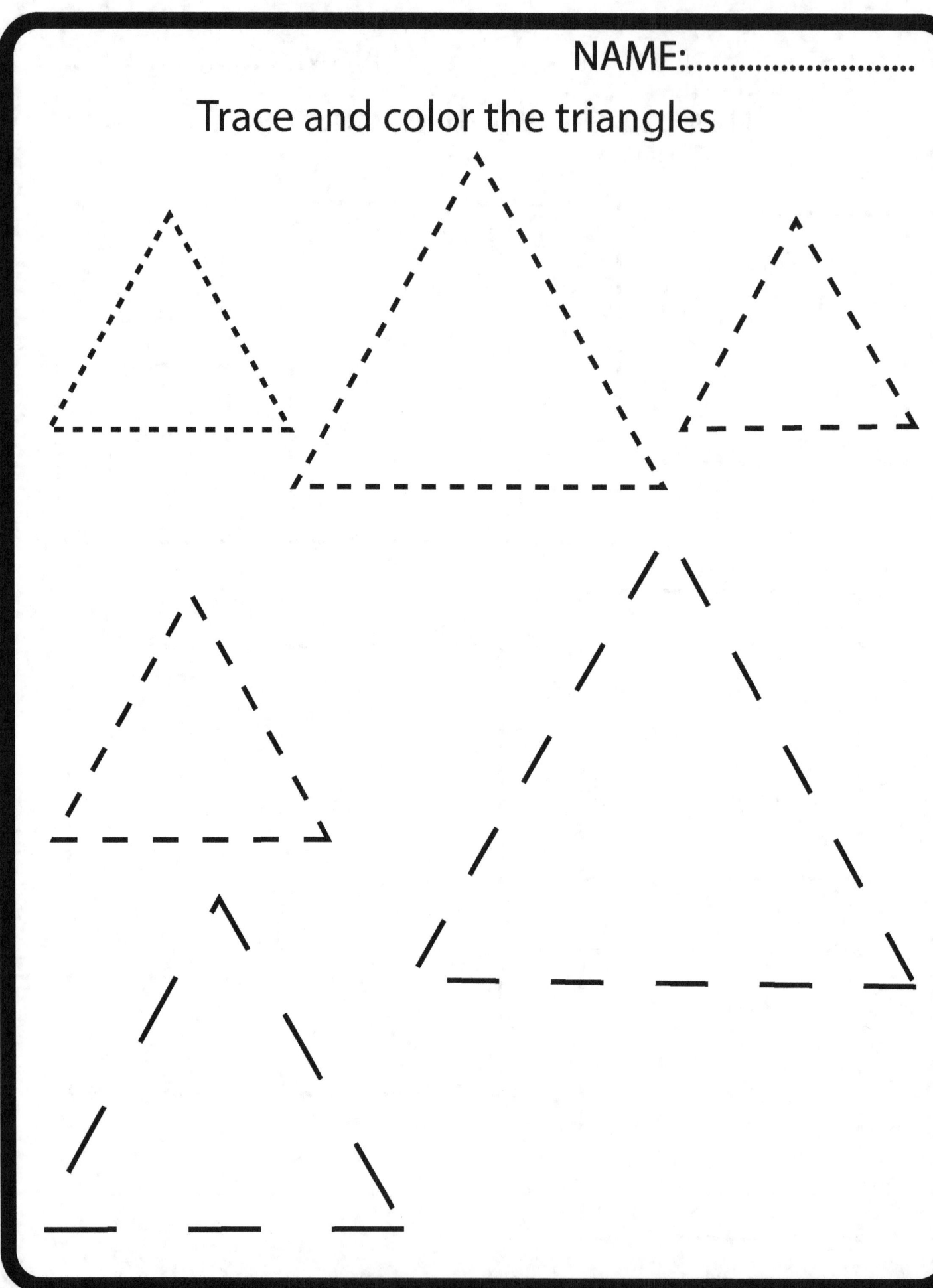

NAME:...........................

Trace and color the ovals

Trace and color the stars

Trace and color the rectangles

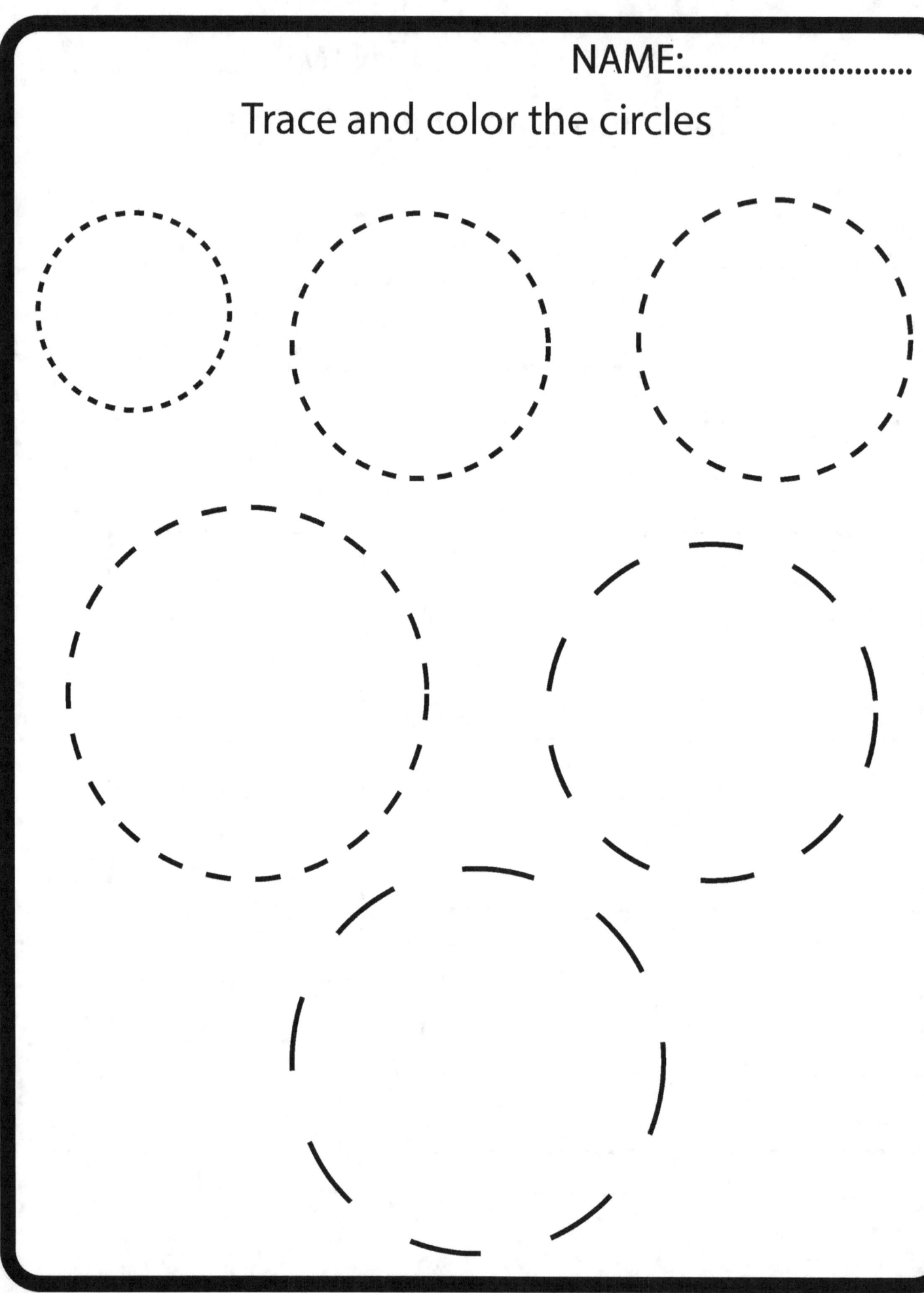

NAME:.........................

Trace and color the circles

Match the shapes and picture below

 •

•

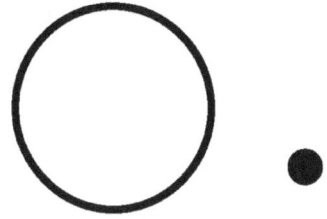 •

•

 •

•

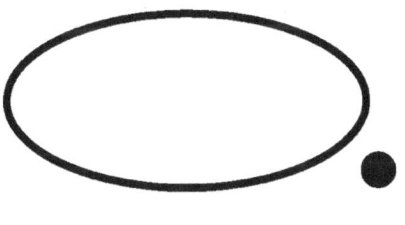 •

•

 •

•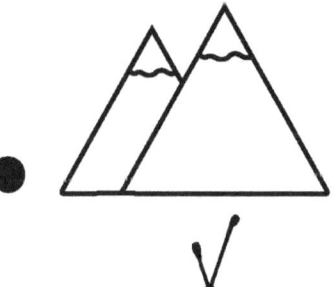

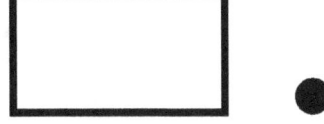 •

•